# MERCURY POLLUTION

## A Transdisciplinary Treatment

# MERCURY POLLUTION

## A Transdisciplinary Treatment

*Edited by*

Sharon L. Zuber and Michael C. Newman

CRC Press
Taylor & Francis Group
Boca Raton London New York

CRC Press is an imprint of the
Taylor & Francis Group, an **informa** business

Cover Insert: Artisanal gold miners using mercury at the El Pache refinery in Porto Velo, Ecuador. (Photo by Kris Lane, 2007.)

CRC Press
Taylor & Francis Group
6000 Broken Sound Parkway NW, Suite 300
Boca Raton, FL 33487-2742



Printed and bound in Great Britain by TJI Digital, Padstow, Cornwall
Version Date: 20110711

International Standard Book Number: 978-1-4398-3384-1 (Hardback)



**Library of Congress Cataloging-in-Publication Data**

Mercury pollution : a transdisciplinary treatment / editors, Sharon L. Zuber and Michael C. Newman.
p. cm.
Summary: "Using mercury pollution as an example, this book illustrates the interdisciplinary approach required for solving environmental problems. It explains the details of the natural cycling of mercury in and on the earth and discusses how humans have dramatically disrupted its exchange among the earths soil, air, and water. The chapters discuss history, media, and politics in relation to mercury and contain links to established websites with specific resources for readers. Also included are smaller case studies, such as the Minamata tragedy, fish consumption, and international treaties"-- Provided by publisher.
Summary: "Mercury is the gravest chemical pollutant problem of our time, and this is the first publication that has undertaken holistic coverage of this truly global issue. This book is an outstanding product of the innovative Global Inquiry Groups (GIG) process of interdisciplinary collaboration, which brings together the physical sciences, social sciences, and the humanities including the study of media, visual, and literary arts, uniquely enabling us to look at and understand mercury from the many perspectives needed in order to grasp the problem in its totality. This book clarifies how we are all connected to mercury, how we take it in through the food we eat and the air we breathe, and how we release it as a consequence of our lifestyle. It tells us about the relationship people have had with mercury from ancient to modern times. It tells us how people have given artistic expression to the ravages of mercury, touching our emotions and changing us. It refers to the social injustice of mercury pollution. This book is an attempt to communicate beyond the walls of academia to a larger audience. Therefore, on behalf of the authors of chapters within this book and the mercury group, I invite you to explore metal mercury from the varied perspectives of history, science, sociology, government, writing, and art. Each chapter represents a disciplinary thread from the GIG. Our goal was to weave these interdisciplinary threads into a tapestry that presents a more complete picture of the effects of mercury pollution and to provide new ways to think about the environment and our individual responsibility toward each other and our earth"-- Provided by publisher.
Includes bibliographical references and index.
ISBN 978-1-4398-3384-1 (hardback)
1. Mercury--Environmental aspects. 2. Mercury--Toxicology. 3. Mercury--Health aspects. I. Zuber, Sharon L. II. Newman, Michael C.

TD196.M38M4686 2011
363.738'4--dc23 2011012189

**Visit the Taylor & Francis Web site at**
**http://www.taylorandfrancis.com**

**and the CRC Press Web site at**
**http://www.crcpress.com**

# Contents

# Foreword

*Aileen M. Smith*

Mercury is the gravest chemical pollutant problem of our time, and this is the first publication that has undertaken holistic coverage of this truly global issue.

This book is an outstanding product of the innovative Global Inquiry Groups (GIG) process of interdisciplinary collaboration, which brings together the physical sciences, social sciences, and the humanities, including the study of media, visual, and literary arts, uniquely enabling us to look at and understand mercury from the many perspectives needed in order to grasp the problem in its totality.

The GIG process is based on international and cross-disciplinary collaborations by students and faculty at the College of William & Mary campus interfacing with researchers, artists, and activists holding various positions—some colliding—on the mercury issue, and subsequent sharing of findings with the community beyond campus through symposia and student presentations (some of them from field trips abroad), art exhibitions, and film screenings. This GIG's work, therefore, also serves as a model of what can potentially be achieved from within one college. It helps us think about how the college campus can aid us in creating essential connections to solve the challenges of our time.

Publication of this book is also very timely. In February 2009, after years of hard work by citizen organizations, victims of mercury pollution, researchers, and scientists, the Governing Council of the United Nations Environment Programme (UNEP) agreed on the need to develop a global, legally binding instrument on mercury. The second session of the Intergovernmental Negotiating Committee to prepare this instrument took place in Japan in January 2011. The goal is to complete all necessary negotiations before the United Nations Governing Council/Global Ministerial Environment Forum of 2013.

This book clarifies how we are all connected to mercury, how we take it in through the food we eat and the air we breathe, and how we release it as a consequence of our lifestyle. It tells us about the relationship people have had with mercury from ancient to modern times. It tells us how people have given artistic expression to the ravages of mercury, touching our emotions and changing us. It refers to the social injustice of mercury pollution. The brunt of the problem has been imposed most on those less economically and socially advantaged: indigenous peoples, artisanal miners, and the poor who live in coastal areas afflicted with mercury pollution. Society and politicians, however, have ignored or paid little attention to their plight. Even today prevention and compensation lag far behind, and the world's most vulnerable populations continue to be disproportionally affected.

The largest contribution by individuals in most industrialized countries comes from their use of electricity, because coal-fired power plants have become the major source of mercury caused by human activity. The myriad uses of mercury in lamps, lightbulbs, and medical applications are also large contributors. Mercury is the only metal found as a gas in the earth's atmosphere. It travels globally. And our land and waters are polluted both locally and globally from uses of products containing mercury.

My personal realization of our interconnectedness in this world came to me quite unexpectedly twenty-five years ago, at the time of the Chernobyl nuclear accident in 1986. In Japan, 8,000 kilometers distant from the explosion, a colleague of mine who is a nuclear physicist was checking his radiation monitoring equipment, thinking it would be impossible for the radioactive cloud to travel as far as Japan. There was no sign of any release from the accident until five days later, when suddenly the display screen reacted wildly. The radioactive cloud had landed. People around me talked about how shocked they were, but in that instant I had a completely different vision. It was not just that week that we were connected to that part of the world. Last year, a decade ago, 100 years ago, and also tomorrow, next year, and thereon, we had been and would remain continually connected. Not only were we connected to that place 8,000 kilometers away, we were connected to everything between there and here, and whatever we did here in Japan we were passing on to other places. Suddenly I realized that even my own body was not alone. It was the sum total of the living history of this earth. The water and everything inside me had been shared before and I was sharing it as I lived and would pass it on when I died.

My partner, the photojournalist W. Eugene Smith, and I first heard about the methylmercury poisoning tragedy of Minamata Disease back in 1971. The chemical company Chisso had polluted the waters surrounding Minamata Bay in Japan. The poisoning was discovered in 1956, when it had reached epidemic proportions. When we decided to go and photograph it, I remember some scientists telling us, "Why go now to photograph? The incident is over. It's history." But I learned subsequently, during the three years Eugene and I lived and photographed there, that this was far from the truth (see Figure 1).

Even today the victims of Minamata Disease, the largest single incident of mercury poisoning caused by industrial pollution, are still fighting for justice in the courts. Lack of action on the part of the Japanese government and industry has resulted in more than 40,000 victims in Japan today who are either recognized as suffering from this methylmercury poisoning or still seeking recognition as Minamata Disease victims. The Japanese government has yet to undertake a proper epidemiological study of the people affected by the contamination and has never supplied any explanation to substantiate the claim that the government's medical criteria for victim certification are based on sound medical science. We remain right in the thick of this historic tragedy.

I tell young people in Japan, "When you see visual images of the contorted limbs of the Minamata victims, do not feel sorry for them. You and they are connected. It is because they and other pollution victims fought in the courts that much stricter environmental laws were enacted in Japan in the mid-1970s." I tell them that, "Because of this, your parents, when you were developing in your mother's womb and when you were growing up, were exposed to much less chemical pollutants. If it weren't for the people you see in these photographs, you yourselves would have that much more chemical pollution in your body today."

History has taught us that reducing the supply, demand, and trade of mercury, rather than trying to control its releases, is the most effective way to deal with this chemical.

It is up to the collective us whether or not humanity will be able to create this vitally necessary global, legally-binding treaty properly and in a timely manner at the UN level, implement it, and initiate and implement controls locally including those which will affect our own lives. We are, indeed, all active players in the global mercury picture. Mercury is a reflection of our interconnectedness. The only way to reduce this chemical hazard is to be aware of this connectedness and, through our awareness, take action.

**Aileen Mioko Smith**
*Coauthor,* Minamata, *with W. Eugene Smith*
*10 September 2010*
*Kyoto, Japan*

**FIGURE 1** Shinobu and her mother preparing to leave the Minamata train station for the United Nations Environmental Conference in Stockholm, Sweden. In *Minamata*, Photograph by W. Eugene Smith, 1972. Courtesy of Aileen Archive.

# Preface

*Sharon L. Zuber*

What do a baby bird, a *Life* magazine photograph, a reddish-colored rock, and blue sky have in common? Each made mercury pollution visible for a transdisciplinary group of faculty members from the College of William & Mary and inspired the collaborations for this book.

The seed for these collaborations was planted in 2006 when Dr. Laurie Koloski, director of the Reves Center for International Studies, in partnership with Joel Schwartz, director of the Roy R. Charles Center for Honors and Interdisciplinary Study, issued an invitation for William & Mary faculty to form Global Inquiry Groups (GIGs) focused on a single research interest. The GIG concept was designed to encourage international and cross-disciplinary collaborations in the areas of undergraduate research, curriculum innovations, and public outreach or, at the very least, to get people "across campus" talking.

Just weeks before the GIG announcement appeared, biology professor Dan Cristol asked me if any of my "Film Studies" students would be interested in videotaping work he was doing about mercury in songbirds on the South River in Waynesboro, Virginia. One of my students, Liz Budrionis, a major in neuroscience and minor in film, was a perfect fit for this project since mercury is a known neurotoxin. Serendipitously, Dr. Li Xiong from Central China Normal University (CCNU) in Wuhan was working on postdoctoral research in ecotoxicology and heavy metals at the Virginia Institute of Marine Science (VIMS), a graduate program at the College of William & Mary. She chose to work with Dr. Michael Newman because of his international reputation in this area and because she had used his *Fundamentals of Ecotoxicology* textbook in a class she taught in China. Her one-year stay in Virginia gave her the opportunity not only to work in the VIMS lab but also immerse herself in US culture, which included sharing a house with me.

Over dinner one evening I began talking about a possible film project in Waynesboro, Virginia, and Li Xiong mentioned that one of Professor Newman's research projects was also on the South River. Soon we were talking about what exactly "ecotoxicology" meant, and Li Xiong explained that she fed toxicants to fish or shrimp not to kill them but ultimately to save them. It quickly became obvious that mercury in the environment linked four faculty members from different disciplines and cultures in a concern about the environment as well as in our research and teaching.

Thus, in January 2007 Dan Cristol, Li Xiong, Mike Newman, and I met to discuss organizing an Exploratory GIG (E-GIG) based on the subject of global mercury pollution. We met at a restaurant where our conversation quickly turned to the mercury bioaccumulation in fish as a way to select the best, safest lunch entré. One goal for the E-GIG was to bring together as many people as possible who had an interest in environmental education. The focus on mercury pollution provided a case study as a way to promote research projects, courses, and public events that would emphasize the global nature of the mercury hazard and how it touches all disciplines with concerns about science, education, economics, law, art, etc. Our vision was to share information about mercury pollution that would be accessible to a general audience, not just scientists and college professors. More specifically our goal was to see if enough people were interested in the topic to apply for a three-year grant that would allow us to expand the interdisciplinary and international connections that had so energized our group.

Throughout the semester the E-GIG sponsored three general meetings and a field trip open to the campus community. Meeting topics focused on "Ecological Health," "The History and Globalization of Mercury," and "Mercury and Human Health." In May the group took a field trip to Waynesboro to see Dan's research project monitoring mercury in songbirds. We saw the factory, once owned by DuPont, that was responsible for the unintentional mercury pollution during the 1950s as a by-product from the manufacture of rayon. We learned about "biomagnification," or how methylmercury accumulates, through natural and chemical processes in fish and wildlife. We held baby birds (nestlings), explored nesting sites, and discussed theories about why the mercury in the South River wasn't dissipating over time as originally predicted and, in fact, seemed to spread far beyond the riverbed. The importance of research at this site was clear; there were few places in the world with such high concentrations of mercury, and what data scientists collected could help researchers around the world who were interested in mercury contamination.

Mike Newman introduced our group to photographs by W. Eugene and Aileen M. Smith that some of us remembered from the covers of *Life* magazine more than thirty years ago. These striking black-and-white images brought the Minamata mercury tragedy into sharper focus for the group. The Smiths traveled to Japan to document the tragedy sparked by the Chisso plant dumping methylated mercury into Minamata Bay. The methylmercury poisoned the fish that provided a livelihood and sustenance for the Japanese community and resulted in thousands of deaths, children born with disabilities, and public health care issues that continue into the twenty-first century. As Mike wrote in his essay for the exhibit catalogue of the Smiths' photographs, these images—more than any scientific or health report—"wrenched our collective attention away from the benefits of giddy industrialization, revealing the blatantly unacceptable human consequences of pollution" (see Chapter 1).

Mike's presentation and the field trip intrigued art professor and sculptor Elizabeth Mead. She was inspired to develop a new course, "Sculpture in the Global Environment," which explored the relationship between sculpture and environmental issues and that led to an art exchange with Japanese art graduate students and student projects from her new course. Her interest in the photographs led to her curating an exhibit of the Smith photographs for the Muscarelle Museum on the William & Mary campus that opened in 2010: *Unbearable Beauty: Triumph of the Human Spirit.* Clearly, the cross-pollination between a scientist's presentation and an artist's sensibilities produced representations of mercury poisoning that found their way into public venues and around the globe. People who had never thought about mercury pollution were creating art that expressed their knowledge and personal connection to a global problem.

Providing the historical context for our project was Professor Kris Lane. He talked about the mercury mines in Spain and passed around the heavy, reddish-colored sample of cinnabar (wrapped safely in a cloth). We learned about mercury use throughout history and about how this "dangerous attraction" still used today has begun to lose its luster in the public imagination. Mercury's continued use in artisanal gold mining in South America raises issues of economic inequality and public health communication in local communities, issues that would benefit from scientific and international collaboration.

Throughout the spring semester many new relationships were forged among faculty who came from a variety of disciplinary backgrounds. In addition links were developed globally through Kris Lane's work in Colombia and Ecuador and Li Xiong's participation from Wuhan, China. She helped sponsor three faculty–student exchanges that led to cross-cultural perspectives and new ways of understanding environmental concerns. The exchanges began with three Chinese students visiting the William & Mary campus in the spring of 2008. When the CCNU students had been in the United States for a few days, I asked them what most impressed them about Virginia—they were unanimous when they replied, "The blue sky!" I only fully understood their amazement after I had the opportunity to travel to Wuhan, a city of more than 8,000,000 people supported by burgeoning industries and filled with throngs of automobiles. The Chinese students were accustomed to a hazy sky and aware of the price their environment and health were paying for fast-paced industrialization. The E-GIG discussions generated enough interest for a core group of faculty to draft a grant proposal for a three-year effort that would include funding for undergraduate research, new courses, and public events. We were no longer exploratory; we became an S-GIG, or Mercury Sustained Global Inquiry Group. With faculty from biology, sociology, art, film studies, geology, environmental science, and history as well as graduate and undergraduate students, our GIG created a transdisciplinary portal through which to explore international environmental hazards and to instill stewardship and appreciation for the environment. The topic of mercury provided an opportunity to explore the social, cultural, economic, and policy implications of global pollution and emphasize that science is not just contained within the border of a laboratory nor are laboratories separate from politics and cultural production.

Over the next three years the energy and enthusiasm produced by this group led to funding student and faculty research projects from "Fetal Exposure to Mercury" to "The Local Dimensions of a Global Hazard: Mercury and Gold Recovery in Southern Ecuador"; new courses such as "Mining the Americas"; and a series of public lectures and film offerings. We learned about the effects of mercury pollution on indigenous peoples from Amy Roe, research assistant at the University of Delaware's Center for Energy and Environmental Policy; Marcello Veiga, from the Department of Mining Engineering at the University of British Columbia; and Robert Davidson, researcher at the Biodome and the Amazon. Filmmaker Hardy Jones focused on the threat to dolphins (anticipating the Oscar winning *The Cove* for which he was interviewed). Dr. Vera Stejskal, associate professor of immunology at First Medical Faculty at Charles University in Prague, Czech Republic, came from Europe to talk about allergic reactions to heavy metals putting at risk children due to vaccines and dental hygienists, who work with fillings that are more mercury than "silver." Pep Fuller, EPA's first standards executive, talked about the difficulty in negotiating a global agreement to curb mercury pollution in spite of the new administration's declaration of mercury as "the world's gravest chemical problem" (Maliti, *Seattle Times*, Feb. 17, 2009).

Three years of hard work and planning culminated in the 2010 International Mercury EXPO, "Mercury: A Hazard without Borders" (see Figure 2). Coordinated with the fortieth anniversary of Earth Day, this weekend EXPO included panels, workshops, exhibits, performances, and field trips. The South River Science Team held its quarterly meeting in Williamsburg so its members could attend the sessions, and Sheryl Telford, director of the DuPont Corporate Remediation Group, was the keynote speaker at the luncheon. The variety of participants and presentations showcased the value of a transdisciplinary approach to this important topic. Attendees came from as far away as Belize, Peru, and Japan. In addition, the events attracted William & Mary students and faculty as well as Williamsburg community members and included a special program for middle school students from local area schools.

Aileen M. Smith flew from Japan to give the opening talk for the Muscarelle Museum exhibit mentioned above. Her talk, "Living and Working with W. Eugene Smith: Integrity, Human, and Scotch," touched on her professional and personal relationship with the famed photographer and her collaboration with him on the Minamata photographs. The Smiths gave a face to the insidious, invisible tragedy, a face that circulated around the globe as a cover for *Life* magazine. As important, the audience learned that the tragedy continues for many Japanese families who are still fighting for recognition and compensation related to lingering effects of Minamata Disease. An environmental activist, Aileen Smith travels the world on behalf of the victims; at the same time she makes an effort to stay in close contact with many of the individuals. Her presentation, although focusing on the photographs and her relationship with W. Eugene Smith, made clear the importance of attention on the issue of mercury pollution, not just as a continuing scientific and health concern but also as an issue of moral responsibility.

During the EXPO the William & Mary Andrews Gallery housed artwork by Lynda Wysong, an artist who also confronts environmental pollution through her sculpture and prints. Participants were free to browse this gallery and the Smith exhibit as well as watch a performance by the Virginia Theater Machine, *A Mercurial Roadshow*, written specifically for the EXPO. By starting a conference about global mercury pollution with a focus on art, we hoped to emphasize the value of interdisciplinary collaboration and of finding methods to translate science and health issues in ways that would be compelling for a general public.

At the end of the conference one participant, an accountant and amateur bird watcher from Northern Virginia, who attended the presentations by Dave Evers, founder of the BioDiversity Research Institute, and other ornithologists commented: "When would I ever have been able to learn from these specialists? I can't wait to get back and share what I have learned." An Asian Studies professor joined the improvisation workshop taught by Lisa Jo Epstein, founder and artistic director of Gas & Electric Arts in Philadelphia, that focused on communicating sensitive issues.

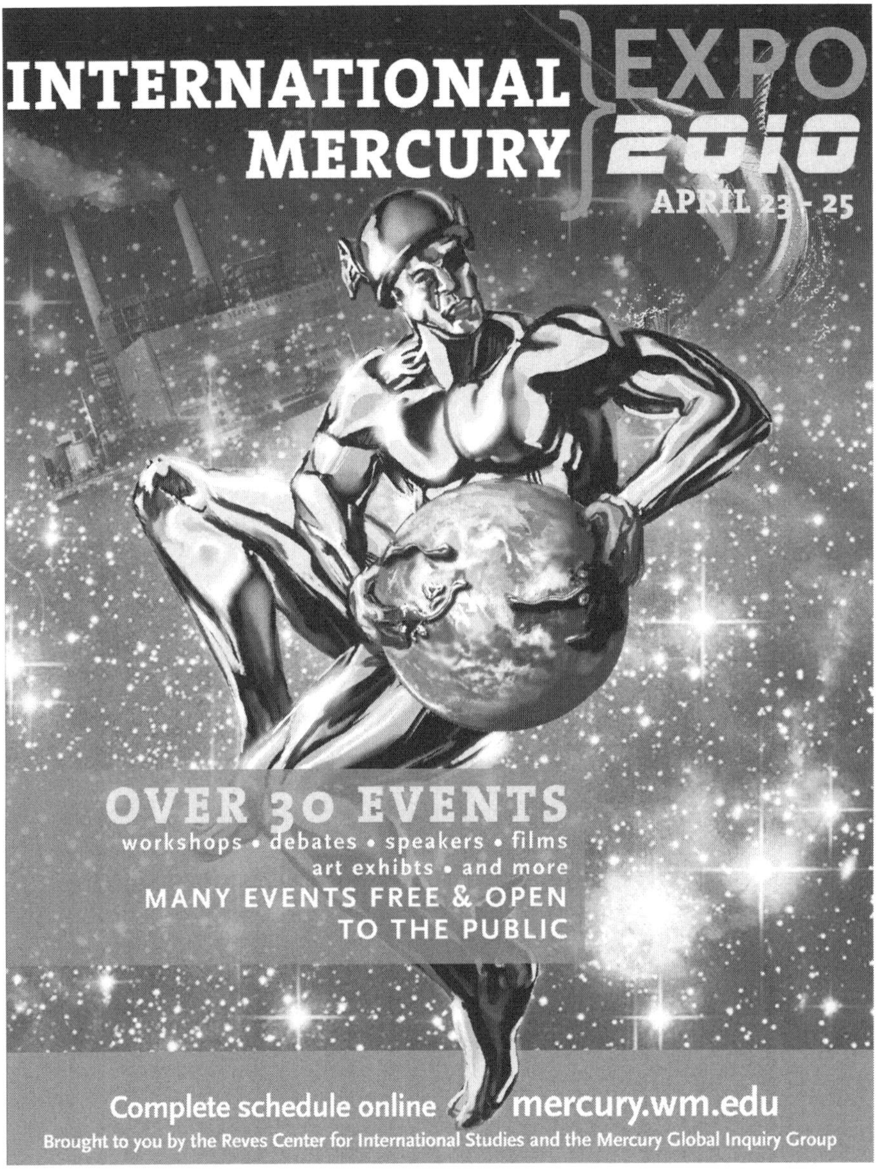


**FIGURE 2** EXPO poster. S. Ryan Lewis ('10), Creative Services, College of William & Mary.

A panel of dentists openly discussed the controversial politics surrounding mercury dental amalgams in light of the American Dental Association's support of their continued use. Above all, the EXPO created a space for discussing mercury pollution, one not restricted by departments, levels of professionalism, or age. The GIG goal, to emphasize the importance of communication across boundaries whether in writing, performing, creating works of art, or teaching, was met through the varied events.

Over the three years leading up to the EXPO the mercury core met frequently, and we found ourselves actually looking forward to meetings. Gathered around a seminar table we came to expect that our discussions would spark new ways of thinking and generate ideas that we could take back to enhance our own research and teaching. At one such meeting we started bantering about the idea for a book that looks at mercury from a variety of disciplinary perspectives, a book that would reach out to an audience of teachers at all educational levels, undergraduates, representatives to community forums, public libraries, and more. Along with the EXPO the book would represent the culmination of our collective efforts to share the inspiration of our interdisciplinary collaborations.

As codirectors of the GIG, Dan Cristol and I symbolized these collaborations: he was the science side whereas I represented the humanities, but we were both willing to cross academic boundaries. Dan was a driving force behind the Virginia Theater Machine's original play and served as consultant for its scientific accuracy. He sets high professional standards for himself and his students in the classroom and lab, yet he willingly put on a fake mustache, hairnet, and apron to assume the persona of a fishmonger for the middle school field trips.

Someone asked me what it had been like working with scientists. My answer: mind-expanding! Not only did I learn scientific concepts—for example, the conditions under which metal mercury becomes methylated and how mercury biomagnifies in the food chain—but also what it means to do field work and how important it is to collect accurate data. When I worked with Liz Budrionis on her video about Dan's research at the South River, we got up at 5:30 a.m. to check bird boxes and videotape Dan and his students collecting blood and feather specimens. I learned that a person could hold a baby bird and return it to its nest without the parents abandoning it, which led to my being able to save three baby birds later in the summer.

In addition, how I look at "science" changed. I learned that scientific results are not always clear-cut but often involve complex risk assessment to make decisions. Although I have recycled for more than thirty years and tried to be energy conscious, I now have a better idea why using compact fluorescent lightbulbs (CFLs), even though they contain mercury, is better than having to build another coal-fired power plant, but we need to support proper recycling of these bulbs; one solution may create other problems to be solved.

I also learned that different disciplines use the same words to mean different things: "interdisciplinary" to a scientist might mean a mixture of chemistry, biology, and geology—all sciences to me—whereas to someone in the humanities "interdisciplinary" might mean a mixture of sociology, biology, literature, art, and geology. When biologists use the word "gender" it means "sex"; whereas in the humanities we make the distinction that "sex" is biological and "gender" is socially constructed. I learned how difficult it is to communicate science to a public audience. Exciting research is taking place in William & Mary labs and in the field by our students and faculty; however, translating this information for nonscientists and the general public is a challenge. So, you might ask, how do we communicate?

This book is an attempt to communicate beyond the walls of academia to a larger audience. Therefore, on behalf of the authors of chapters within this book and the mercury group, I invite you to explore metal mercury from the varied perspectives of history, science, sociology, government, writing, and art. Each chapter represents a disciplinary thread from the GIG. Our goal was to weave these transdisciplinary threads into a tapestry that presents a more complete picture of the effects of mercury pollution and to provide new ways to think about the environment and our individual responsibility toward each other and our earth.

# Acknowledgments

Without the Mercury Global Inquiry Group (GIG), this book would not exist. Many thanks go to Dr. Laurie Koloski for the GIG concept, and the Wendy and Emery Reves Center, including Judy Davis and Amy Keuneker. The Mercury GIG core participants deserve accolades: Vivian Appeler, Dan Cristol, John Drummond, Monica Griffin, Erica Holloman, Kelly Joyce, Jim Kaste, Kris Lane, Elizabeth Malcolm, Elizabeth Mead, Mike Newman, and Li Xiong (many of these names appear as chapter authors). At the same time they were teaching and researching full time, these people worked tirelessly to bring the topic of mercury pollution into an open forum.

On board for specific projects were the students who worked at our sides: Liz Budrionis, Kevin Carey, Kathy Eggers, Shelley Holder, and Ashley Pierce. Other groups and individuals contributed to the EXPO and to spreading the word about mercury pollution to a general audience: the William & Mary Student Environmental Action Coalition (SEAC); the Committee on Sustainability (COS); the Improvisational Theatre Group; Marie Flowers and Linda Cifelli from Dental Amalgam Mercury Solutions (DAMS); the Muscarelle Museum's Director Aaron De Groft along with Amy Gorman, coordinator of education and media, and Ursula McLaughlin-Miller, special projects administrator; and William & Mary President Taylor Reveley and the College of William & Mary administration.

Special thanks go to Adam Stackhouse, without whom many of the GIG events, including the EXPO, would not have happened. In his own ranging curiosity, creativity, ability to coordinate and communicate with seemingly disparate interest groups, not to mention his videography and organizational skills, Adam represents the heart of the Mercury GIG ways of thinking and acting that transcend traditional borders.

Artwork, in the form of the vibrant poster for the EXPO, as well as the maps, mercury cycle, and icons in this book, were drawn by S. Ryan Lewis ('10), working at William & Mary Creative Services (see Figures 2, 3, and 4). Elizabeth Mead deserves special recognition for her help obtaining permission to publish photographs originally used in the Muscarelle Museum exhibit that she curated as well as her artistic collaboration on the cover design.

Finally, special thanks to Aileen M. Smith for permission to publish photographs taken by W. Eugene Smith and herself made available through the Aileen Archives with the help of Masako Hino. We hope that this book not only will keep attention on but also open up ways of thinking about the problems of global mercury pollution to make our world a safer place to live and work.

# The Editors

As codirector of the Mercury Global Inquiry Group, **Sharon L. Zuber** organized faculty and student exchanges with Central China Normal University and the 2010 International Mercury EXPO, as well as mentored undergraduate student research video projects about mercury pollution. In 2006 she was invited to speak at a Symposium on Teaching and Research in Academic Writing at Waseda University in Tokyo, Japan. In 2008 she was area chair for "Environmental Documentaries" for the Film & History Conference. She teaches writing-intensive courses and works with faculty across the curriculum teaching freshman seminars.

Her research interests focus on nonfiction writing and documentary filmmaking and include a comparison between New Journalist writers and Direct Cinema filmmakers, movements that emerged in the 1960s in the United States Her documentary films include *They Live in Guinea* (1996), a video about the Gloucester, Virginia, watermen; *Master of the Flame* (2004), a documentary about glass artist Emilio Santini; *Mercury on the Move* (2008), a short about the research about mercury in songbirds by Professor Daniel Cristol; and *Mercury: A Hazard without Borders* (2009), a film about the history and issues surrounding mercury pollution. She has published in *College English*, *Post Script*, and *Film & History*.

**Sharon L. Zuber**
*Director, Writing Resources Center*
*Assistant Professor, English Department and Film Studies Program*
*College of William & Mary*
*Williamsburg, Virginia*

**Michael C. Newman** served as dean of graduate studies in the College of William & Mary's School of Marine Science from 1999–2002. Previously he was a faculty member at the University of Georgia's Savannah River Ecology Laboratory. His research interests include quantitative ecotoxicology, environmental statistics, risk assessment, population effects of contaminants, metal chemistry, and bioaccumulation and biomagnification modeling. In addition to more than 115 articles, he authored five books and edited another five books on these topics. A 2006 Mandarin translation of his *Fundamentals of Ecotoxicology* is available (Chemical Industry Press, Beijing), and a Mandarin translation of his marine risk assessment book will appear in 2011.

He taught at universities throughout the world including the University of California—San Diego, University of South Carolina, University of Georgia, College of William & Mary, Jagiellonian University (Poland), University of Antwerp (Belgium), University of Joensuu (Finland), University of Technology—Sydney (Australia), University of Hong Kong, University of Koblenz-Landau (Germany), University of Xiamen (China), Huazhong Normal University (China), and Royal

Holloway University of London (UK). He served numerous international, national, and regional organizations including the Hong Kong Areas of Excellence Committee, OECD, US EPA Science Advisory Board, US EPA ECOFRAM, US EPA STAA, and the US National Academy of Science NRC. In 2004 the Society of Environmental Toxicology and Chemistry awarded him its Founder's Award, "the highest SETAC award, given to a person with an outstanding career who has made a clearly identifiable contribution in the environmental sciences."

**Michael C. Newman**
*A. Marshall Acuff Jr. Professor of Marine Science*
*School of Marine Science*
*College of William & Mary*
*Williamsburg, Virginia*

# Contributors

**Sarah Jane Brubaker** teaches undergraduate and graduate courses in gender, gender and the body, medical sociology, theorizing gender violence, and sociological theory. Brubaker conducts research in the areas of adolescent sexuality and pregnancy, especially among African-American girls; girls and juvenile justice; gender violence; and medicalization. She specializes in qualitative research methods and program evaluation. Her research has been published in the *Journal of Marriage and Family*, *Gender & Society*, and *Feminist Criminology*.

**Sarah Jane Brubaker**
*Associate Professor and Graduate Coordinator of Sociology*
*L. Douglas Wilder School of Government and Public Affairs*
*Virginia Commonwealth University*
*Richmond, Virginia*

**John G. Drummond** became part of the college's Mercury GIG through his academic background in the humanities, vocation in technology, and good fortune.

**John G. Drummond**
*Lead Engineer for Academic Information Services and Adjunct Instructor*
*College of William & Mary*
*Williamsburg, Virginia*

**Irving L. "Pep" Fuller** served as principal EPA negotiator for the world's two major commercial chemicals treaties: The Rotterdam Convention on Prior Informed Consent for Banned and Severely Restricted Chemicals (PIC), and the Stockholm Convention to Phase Out the Use of Persistent Organic Pollutants (POPs). For these and other achievements President Clinton conferred on him the rank of Meritorious Senior Executive. In 2001 he retired from government service including the United States Marine Corps, the Foreign Service, EPA and the US Trade Representative's Office, Executive Office of the President. He lives in Deltaville, Virginia, with his wife Whitney and remains active in international trade and environment issues.

**Irving L. "Pep" Fuller, BA, JD, LLM**
*Deltaville, Virgina*

**Justin Jesty** has a PhD from the University of Chicago and researches modern Japanese literature, film, and culture.

**Justin Jesty**
*Seattle, Washington*

**Kelly Joyce's** research focuses on the sociological dimensions of scientific knowledge and science policy. Joyce's publications include *Magnetic Appeal: MRI and the Myth of Transparency* (2008) and *Technogenarians: Studying Health and Illness Through an Ageing, Science, and Technology Lens* (2010).

**Kelly Joyce**
*Associate Professor of Sociology*
*College of William & Mary*
*Williamsburg, Virgina*

Since the early 1990s, **Kris Lane** has been researching and writing about the history of mining in the Americas. His most recent book is *Colour of Paradise: The Emerald in the Age of Gunpowder Empires* (Yale University Press, 2010).

**Kris Lane**
*Professor of History*
*College of William & Mary*
*Williamsburg, Virginia*

**Kenneth M.Y. Leung** is an aquatic ecotoxicologist with sound knowledge in biostatistics and specialization in ecological risk assessment (ERA). Since 1999 he has published more than eighty peer-reviewed articles that are principally related to the ecology, pollution, ecotoxicology, and ERA in both marine and freshwater ecosystems. Professor Leung was recognized as one of "Ten Outstanding Young Persons" for Hong Kong in 2010.

**Kenneth M.Y. Leung**
*Associate Professor*
*School of Biological Sciences*
*University of Hong Kong*
*Pokfulam, Hong Kong*

**Elizabeth Malcolm's** primary research interests are the transport and transformation of metals in the environment, with a special interest in mercury. She has been studying mercury for more than ten years, beginning as an undergraduate at Earlham College and continuing through graduate school at the University of Michigan and a postdoctorate at Princeton University.

**Elizabeth Malcolm**
*Associate Professor*
*Earth and Environmental Science Department*
*Virginia Wesleyan College*
*Norfolk, Virginia*

**Elizabeth Mead** holds an MFA from Southern Methodist University. Her sculptures and drawings have been exhibited across the United States and in Australia, England, Iceland, Japan, Korea, Portugal. and Taiwan. A recipient of the Japan/US NEA Creative Artist Fellowship, Mead spent six months in Japan where she explored space and dwelling as cultural identifiers. She was recently awarded a Nes Artist Residency in Skagaströnd, Iceland, where she examined the ways environmental cues connect us to the world. She collaborates in a team-taught course about complex systems theory, network dynamics, and sculptural theory and practice. Mead received the William & Mary Alumni Teaching Fellowship Award in 2010.

**Elizabeth Mead**
*Associate Professor of Art and Art History*
*College of William & Mary*
*Williamsburg, Virginia*

**Christine E. Mowery** has a PhD from Vanderbilt University. Her teaching and research interests include gender, social movements, aging and sexuality. She has a forthcoming co-authored chapter, "Exploring Embodied Aging and Ageism Among Old Lesbians and Gay Men" in an edited book on sexualities over the life course.

**Christine E. Mowery**
*Visiting Professor of Sociology*
*University of Richmond*
*Richmond, Virginia*

**Kira Obolensky** is an award-winning playwright and published author of nonfiction books about architecture, including coauthor of the national bestseller *The Not So Big House*. Her newest play, *Cabinet of Wonders: An Impossible History*, is a collaboration with visual artist Irve Dell. *Raskol*, her adaptation of *Crime and Punishment* for Ten Thousand Things Theatre, is on two critics' top ten lists for 2009; and her novella, *The Anarchists Float to St. Louis*, won the 2009 Quarterly West Novella Contest, judged by Padgett Powell. Obolensky has worked collaboratively with choreographers and visual artists and is cofounder of The Gymnasium, a consortium of nationally known artists, scientists, and innovators involved in the incubation of new work and ideas. She teaches at Goddard's MFA Interdisciplinary Arts Program and at the University of Minnesota.

**Kira Obolensky**
*Goddard College*
*Plainfield, Vermont*

**Sydney Landon Plum** is the author of *Solitary Goose*, a collection of nature essays published by the University of Georgia Press in 2007 and nominated for an Orion Reader's Choice Book Award. Plum has been an adjunct in the English department at the University of Connecticut, Storrs, for the past decade, and in 2009 received an award for outstanding teaching. Her poetry and nonfiction have been published in *Prairie Schooner, South Dakota Review, Organization and Environment, ISLE*, and elsewhere. She edited *Coming Through the Swamp: The Nature Writings of Gene Stratton Porter.* Time spent with loons on a lake in mid-coast Maine made Plum aware of the dangers of methylmercury in the environment.

**Sydney Landon Plum**
*Adjunct Professor*
*University of Connecticut, Storrs*
*Storrs, Connecticut*

**Clare Stankwitz** became interested in the international politics of mercury pollution when she attended the twenty-fifth meeting of the UN Environment Programme's Governing Council/Global Ministerial Environmental Forum, where world leaders first agreed to develop an international agreement on limiting anthropogenic mercury pollution. She attended this meeting as part of research through the Global Environmental Governance Project, under the direction of Dr. Maria Ivanova. Stankwitz then took a seminar at the College of William & Mary on heavy-metal contaminants and concurrently completed research on the roles of the three branches of the US federal government in US international environmental policy. Stankwitz ('11) double majored in geology and environmental policy at the College of William & Mary.

**Clare Stankwitz**
*Clifton, Virginia*

**Li Xiong's** work focuses mainly on the toxic effects and the mechanism of heavy metals and insecticides on aquatic organisms such as algae, daphnia, and fish. She has published more than thirty peer-reviewed articles that are principally related to the ecotoxicology, risk assessment, and pollution in freshwater ecosystems.

**Li Xiong**
*Associate Professor*
*College of Life Sciences*
*Central China Normal University*
*Wuhan, China*

# Key to Maps and Related Chapters

Although separated by discipline-specific methods, the chapters in this book are thematically connected by the global nature of mercury pollution. The attached maps of the earth's surface (Figures 3 and 4) attempt to identify the most prominent themes shared among chapters.

 Natural Sources Including Volcanoes, Forest Fires—Along with geothermal emissions and the slow weathering of rocks, volcanoes and forest fires are ways that mercury is naturally released into the environment. Chapters 2, 3, 5, 6, 7, 9, 10, 11.

 Industrial Processes—Ever since the Industrial Revolution, humans have released more mercury into the environment than natural processes. Chapters 1, 2, 3, 6, 7, 8, 9, 10, and Foreword.

 Compact Fluorescent Bulbs (CFLs)—These bulbs contain mercury but reduce overall mercury omissions by decreasing the need for more coal-fired power plants. Chapter 3 and Preface.

 Mining—Whether mining cinnabar, the common mercury ore, or using liquid mercury in the process of refining precious metals, miners around the world are exposed to mercury and its toxic effects. Chapters 1, 2, 3, 4, 5, 6, 7, 11.

 Tuna, Swordfish, Whale, Dolphin—High in the food chain, these creatures have high concentrations of mercury in their meat through biomagnification. Chapters 3, 4, 5, 6, 7, 11.

 Dentist—Dental amalgams contain mercury and have historically put hygienists as well as dentists at risk, especially if they have metal allergies. Chapters 2, 3, 4, 6, 10, and Preface.

Pregnant Woman—The Environmental Protection Agency and Food and Drug Administration recommend that pregnant women and small children avoid eating fish that are high in mercury. Mercury can be transferred to the fetus, causing neurological harm. Chapters 3, 4, 5, 6, 7, 11.

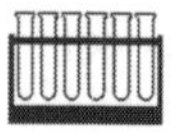

Test Tube, Computer—Careful research about mercury helps to accurately assess its risks. Chapters 4, 5, 6, 7, 8, 10, 11, and Foreword.

Cameras—Through still photos, documentary films, and artistic renderings mercury pollution is made visible for a larger, public audience. Chapters 1, 8, 9, and Foreword.

Global Treaty—Because mercury is a "hazard without borders," all nations are working together to define its uses. Chapter 11 and Foreword.

Writing, Computer, *USA Today*—Without good communication the significance of mercury research would be understood only by scientists and technologists. Reliable reporting of data is crucial to making the public aware of problems as well as informing lawmakers, politicians, and the public. Chapters 6, 7, 10.

Mad Hatter—During the process of curing animal pelts in the 1800s, workers in the hat-making industry were exposed to toxic mercury fumes, leading to illness and even insanity. This and other industrial exposures caused many workplace poisonings across history. Chapters 1, 2, 9, 10.

Polar Bear, Loon, Toucan—Whether far away from coal factories like the polar bear or nesting on lakes and rivers containing methylmercury, wildlife around the globe are affected by mercury pollution. Chapters 3, 4, 10.

People Fishing, Sushi, Rice Paddy—Humans can be exposed to mercury from the food they eat, especially finfish high in the food web. Chapters 1, 3, 4, 5, 8, 10.

 Scales of Justice—The legal system determines what acceptable risk is, establishes environmental laws, and legally links mercury pollution and health issues. Chapters 1, 8, 11, and Foreword.

 Kyoto Arch—Brick-red cinnabar, mercury ore, was used in architecture, jewelry, alchemy, and medicine since ancient times. Chapters 2 and 3.

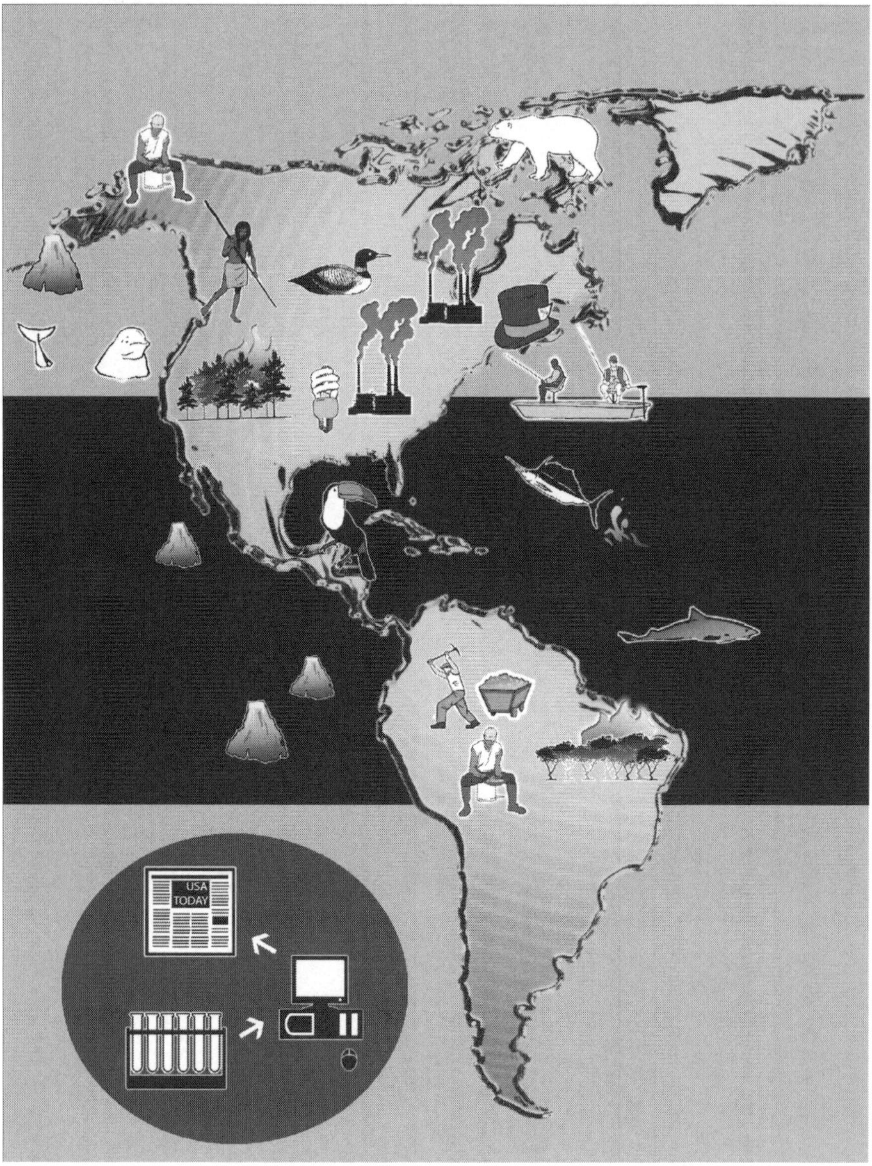


**FIGURE 3** Mercury-related activities in North and South America. The circle highlights the importance of communicating scientific research to the general public. S. Ryan Lewis ('10), Creative Services, College of William & Mary.

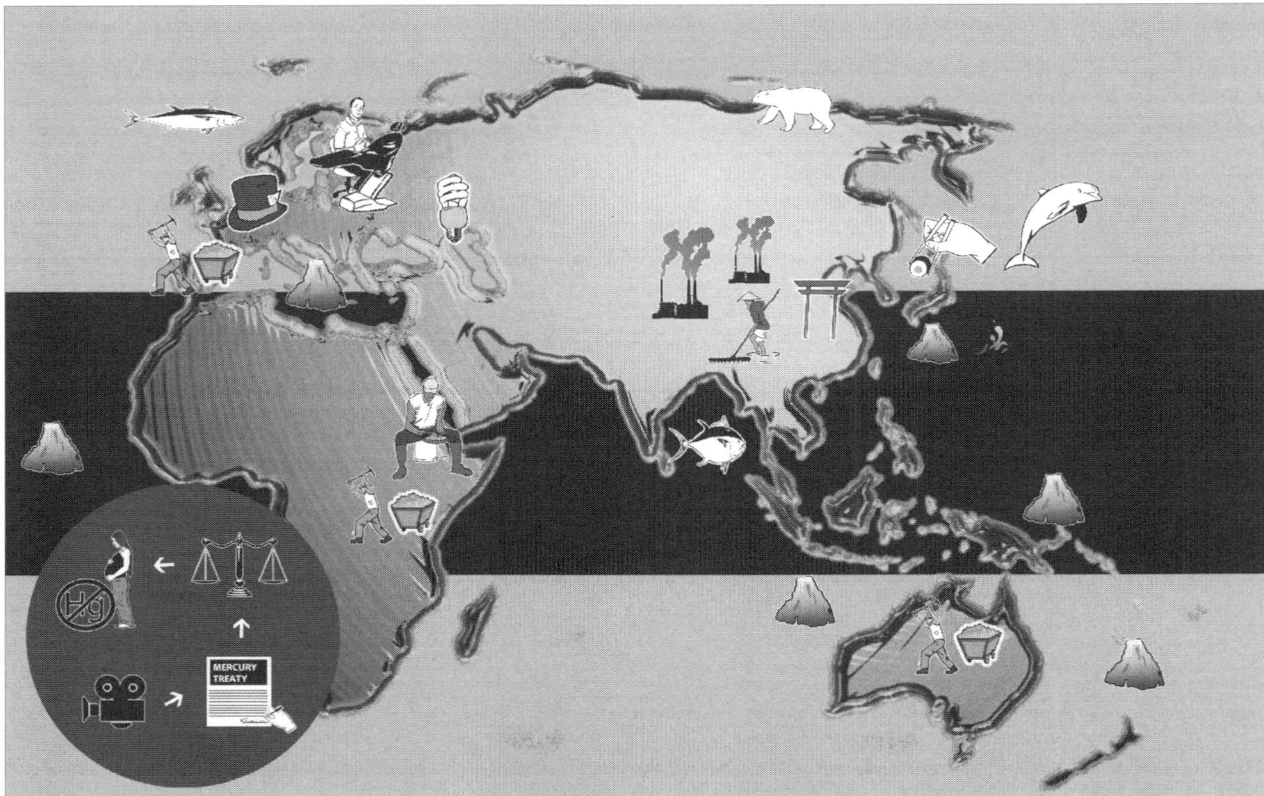


**FIGURE 4** Global activities and objects related to mercury pollution. The circle highlights how the arts, through film, can connect with a wide audience the need for global cooperation and can influence laws that might protect unborn children. S. Ryan Lewis ('10), Creative Services, College of William & Mary.

# 1 When the Scientific Vantage Is Not Enough*

*Michael C. Newman*

## CONTENTS



> Superficiality to me is untruth....
>
> —**W. Eugene Smith,** *Minamata*

## INTRODUCTION

Initially intending to write an introductory chapter for this book about our collective experience with mercury from a scientist's vantage, the familiar way for me to proceed would be to explain the ease with which mercury atoms share outer orbital electrons with other elements. This characteristic makes conversion of mercury to methylmercury easy for sulfate-reducing bacteria in sediments. The typical scientific treatment would next explain how readily methylmercury enters organisms and increases in concentration from one creature to the next in a food web until it reaches harmful concentrations in predators. Referring back to the ease with which mercury shares electrons, I would point out that this same tendency gives rise to damage in tissues, especially nervous tissues, after mercury enters the body. The resulting essay would lay out facts and details while avoiding the scientific sin of subjectivity; it would also fail wholly to describe our human experience with mercury.

What vantage might I take as a scientist and educator to convey meaningful insight if not the comfortable one involving electrons, covalent bonds, oxidative stress, and neurotoxicity? One possible, but unfamiliar, approach suggested itself

* An earlier version of this article originally was published in the catalogue, *Unbearable Beauty: Triumph of the Human Spirit. Photographs by W. Eugene Smith and Aileen M. Smith*, on the occasion of the exhibition by the same name held at the Muscarelle Museum of Art at The College of William & Mary in Virginia, April 24–June 20, 2010.

**FIGURE 1.1** The Cape du Couedic Lighthouse, Kangaroo Island, Australia (M.C. Newman).

after a personal experience. Recently my wife and I took a tour of Australia, including a cold, rainy day trip to Kangaroo Island on the southern side of the continent. One point of interest was the Cape du Couedic Lighthouse (Figure 1.1).

Built in 1909 on an isolated point of the island, this lighthouse became home for a succession of lighthouse keepers and their families. Dangerous surf and sea cliffs restricted visits, so food and essentials for keeper families were hoisted up the sea cliffs by rope and pulley every three months. Each night keepers tended the light, diligently polishing kerosene soot from its optics—optics that revolved smoothly above a reflective pool of mercury. Gossip accumulated about entire families going slowly insane from loneliness at the lighthouse. Scientific analyses have since shown that, while these families might have been lonely, the cause of the insanity was likely mercury poisoning. Staring through the rain at the Cape du Couedic Lighthouse, I tried to imagine the prolonged suffering that must have emerged in each isolated family as father, mother, and perhaps even elder children suffered the disorientation, paranoia, convulsions, and eventual dementia typical of mercury poisoning. This suffering was occurring simultaneously at the Cape Willoughby Lighthouse, elsewhere on Kangaroo Island. Indeed, similar mercury poisonings were occurring in all lighthouses using that particular Fresnel lens design. Objective speculation about mercury's binding affinities and neurological effects was not this scientist's first response. Such a line of thought would have utterly missed the significance of this human experience.

The remarkable features that make mercury so useful—and poisonous—have given rise to many stories laid out in rich objective detail. Historical data demonstrate that our use of mercury dates back for millennia (Martinez-Cortizas et al., 1999). For example, mercury mines in Spain were active during the Roman Empire and served as a brutal means of punishing political prisoners. Convicts in the mines died of mercury poisoning within a year or two of arrival. Sir Isaac Newton's 1693 "year of lunacy" can be attributed to his habit of tasting mercury-rich alchemy concoctions (Broad, 1981). More recent stories provide details about the 1971–72 accidental poisoning of thousands of Iraqis, including 459 deaths, with methylmercury fungicide-coated wheat grains (Bakir et al., 1973) and the ongoing poisonings in the Brazilian Amazon from mercury used to mine gold (Malm, 1998). Presentations of facts and numbers impart accurate medical, epidemiological, or historical insight, but the human experience remains obscured in nearly all of these reports. A solely objective detailing of events cannot portray the anguish, confusion, and painful deaths of people experiencing mercury poisoning. An advocate for objectivity in science, I could only conclude that superficial objectivity was a worse sin than subjectivity in matters such as human poisonings. It would be absurd to construct a typical scientific treatment for the subject as an opening chapter of this book. Instead I choose to illustrate the influence of a nonscientific vantage by describing the extraordinarily subjective attempt of W. Eugene Smith and Aileen M. Smith (see Foreword) to communicate a watershed human experience with mercury.

## W. EUGENE SMITH

W. Eugene Smith (1918–78) came of age professionally while photographing humans caught on World War II Pacific island battlefields. His young Midwesterner's world view broadened at a typical rate during his early years as a photographer in New York City, but it expanded abruptly during his wartime experience, giving rise to the extraordinary empathy so obvious in his images from that point onward. As a photojournalist he revealed the aftermath of island battles and premature deaths of young Navy pilots. His twenty-fifth birthday was spent on the carrier, *Bunker Hill*, where he wrote in his diary about his transformation, "My life, from the moment of my beginning photography, should have veered from much of the superfluous that I poorly recorded, and should have been dedicated to making one part of a record of the tragically momentous years dating from around 1930. I should have religiously spent my entire effort on letting deep history fall before my lens" (Hughes, 1989).* On May 22, 1945, his wartime assignment ended in Okinawa when he was wounded by mortar shrapnel. True to form, the mortar blast was captured by his camera. He spent more than a year undergoing painful surgeries and physical therapy on his hand, face, and palate. The famous photograph of his children taken while he convalesced at his Croton-on-Hudson, New York, home demonstrated that his professional transformation fully embraced the humane as well as inhumane. Taken with a still painful and scarred camera hand, it marked his post-war return to an avocation—and his profound transformation. The innocent confidence captured in *The Walk to Paradise Garden*[†] became the capstone of the renowned *Family of Man* collection (Steichen, 2003) and remains arguably the image for which he is best known.

* Unfortunately, this new passion was not the only feature he integrated into his life from the Pacific theater. He began drinking and taking amphetamines (Hughes, 1989). The alcohol remained, and Benzedrine was eventually replaced by Ritalin as he moved through life and into his final Minamata project.

† The title for this photograph is often misunderstood to refer to the name of a New York garden in which it was taken. It is the title of an intermezzo from "A Village Romeo and Juliet" (Fredrick Delius, 1907) of which Smith was very fond (Hughes, 1989).

Smith's compulsion was now to depict people responding to inhumane or dehumanizing events of the twentieth century. His post-war projects included photographic exposés about African-Americans of the rural South, industrialized Pittsburgh, New York City dwellers, Albert Schweitzer, and finally victims of industrial poisoning. Some projects celebrated the humane, such as his exposés about Maude Callen, a remarkable nurse–midwife serving the poor of rural South Carolina, and the simple grace of Spanish villagers living under Franco's Fascist regime. Other projects captured starker, yet ultimately affirmative, human responses to the inhumane. During his Pittsburgh project he refined his already remarkable approach by participating actively in the depicted story (Hughes, 1989). In his own words, he was unwilling to "remain merely a 'seeing' photographer." Nowhere was this clearer than in his last project (1971–73) that dealt with the accidental mercury poisoning of Japanese villagers. The Minamata project, a joint undertaking with his wife, is the watershed incident used in this opening chapter to explore our human experience with mercury.

## MINAMATA BAY

The people of Minamata lived for generations as a small fishing and farming community on Kyushu Island. They fished Minamata Bay each day as their community grew from town (1912) to small city (1949) (Figure 1.2).

Industrial activity also emerged in Minamata with Nippon Nitrogen Fertilizer Company (later renamed Chisso Company) beginning acetaldehyde production by

**FIGURE 1.2** *Fishing in Minamata Bay*, ca. 1972 (W. Eugene Smith, Center for Creative Photography, University of Arizona: W. Eugene Smith Archive/Gift of Aileen M. Smith).

mercury catalysis in 1932. Typical of the times, Chisso released waste mercury into Hyakken Harbor where it found its way to Minamata Bay. Industry and fishing continued side by side in apparent harmony until 1956. The solution to pollution was dilution.

The first signs that something peculiar was occurring came from Minamata's cats. Cats were acting oddly, displaying loping gaits. Some appeared unhinged and leapt into the sea to drown as apparent suicides. Folks gossiped about the "strange" or "waltzing cat" disease as Minamata's cats became scarce. This curious collection of events warranting only quizzical gossip then coalesced into human tragedy. In 1956, young children began arriving at hospital with symptoms of severe neurological damage. Some were unable to walk or to speak coherently. Others convulsed uncontrollably. Explanations were desperately sought as the number of stricken children increased. Uncertainty about how contagious the disease was prompted ostracism of families with afflicted children, adding to their misery. The Smiths (Smith and Smith, 1975) relate a typical instance of ostracism of one couple as they returned home from the Minamata hospital:

> They had been denied all other means of transportation. They walked along the railroad tracks. ... Mr. Egoshita walked a few steps ahead. On his back he carried the autopsied body of his daughter.

That same year, Dr. H. Hosokawa, a medical doctor working for Chisso, linked a fish-rich diet to the disease and also conducted experiments with cats that implicated waste mercury from Chisso as a possible cause. Soon sales of fish from the area were restricted, curtailing generations of a fishing livelihood for many families. Suspicions about the role of Chisso's mercury waste in this outbreak created conflict between those vested in fishing and those employed by the Chisso Company. Conflict eased for a period when Chisso Company paid modest compensation to a small group of those afflicted, but a similar mercury poisoning event in Niigata stirred the growing number of diagnosed Minamata victims back to civil action. By 1968 mercury released by Chisso and consequent methylmercury accumulation in seafood was identified officially as the cause of what became known worldwide as Minamata Disease. As activists, lawyers, and government regulators began to dominate the process, families were left to cope with the human dimensions of this industrial "mistake."

In 1971 Eugene Smith and his soon-to-be wife, Aileen Mioko Sprague, left their New York loft after eviction notification. By that time Smith's compulsive behavior and addictions had created an intermeshed series of successes and tragedies in his career, health, and personal life. Their plans to visit Japan seemed an attempt to leave behind an overburden of professional and personal issues (Hughes, 1989). Eugene and Aileen went to Japan to hang the photography exhibit, *Let the Truth Be the Prejudice*, in a prominent Tokyo department store. While they prepared to leave for Japan, Mr. Kazuhiko Motomura suggested that they consider photographing the human tragedy unfolding at that time in Minamata. Desirous of a new life and focus for Eugene's energy, the nearly destitute Smiths rented a dirt-floored Minamata house and joined the fisher community (Hughes, 1989). Eugene and Aileen spent what was to become three years documenting the anguish and courage of the Minamata victims. They did so fully engaged in the community, even caring for young victims during brief periods of parental absence.

## MINAMATA'S VICTIMS

The first and most obvious victims were children poisoned in their mother's womb. Mercury in seafood eaten by mothers had passed across the placenta and into cells destined to become nerves of the developing children. Hundreds of babies were born with irreparable damage to their nervous systems. Eugene Smith recollected about their first visit to one young victim's home "We were driving to the village of Akasaki, and dreams flickered in my mind of buying a home here. The illusion ended when we entered the home of Takako Isayama, who was born wrenched from normality, condemned to convulsions and to helplessness" (Smith and Smith, 1975; Figure 1.3).*

As Takako lay "twisted on the floor," the Smiths listened to her parents explain the legal nuance according to which their daughter might not be considered a victim because she was born in 1961. The official stance locked the number of victims at the first 121 diagnosed by setting 1960 as the year that the poisonings ended.† Their daughter was placed onto the unofficial list of those referred to as the lost victims (Smith and Smith, 1975). It is difficult to imagine the emotions that her parents must have felt as they dealt simultaneously with such legal absurdities and the unrelenting needs of their daughter.

Although congenital mercury poisoning was the first acknowledged manifestation of the outbreak, young and old also experienced mercury's impact. Methylmercury poisoned Minamata residents by insidiously destroying already formed nervous tissue including that of the brain itself. Poisoning occurred at a slower pace but with similarly destructive consequences. Mature brain cells were altered irreparably. Ion transport mechanisms crucial for nerve impulses of mature brain cells were damaged. The molecules of which nerve cells are made were destroyed by oxyradicals generated by mercury. Mercury even confused the normal signaling within cells, causing many nerve cells—like Minamata's cats—to commit suicide (Reardon and Bhat, 2007).

Nerve damage and cell death manifested themselves in many ways including muscle weakness, tremors, inability to walk or move normally, damage to vision and hearing, disorientation, and death. The Smiths visited once-healthy young children who became what they referred to as "still-born adults." Mature adults became afflicted too. Yahei and Natsue Ikeda, shown in a series of the Smiths' 1972 photographs (Figure 1.4), were designated official Minamata victims in 1971.

---

* Eugene Smith himself was attacked physically and temporarily blinded by angry factory workers for his intense persistence in depicting the tragic truths of the Minamata victims. The last period of his time in Minamata was filled with his own physical pain due to these injuries. Aileen quickly learned to capture remarkable images when it became too painful for Eugene to operate his cameras (Hughes, 1989).

† A water treatment "Cyclator" was installed by Chisso in 1959, although it is questionable whether it was functional at installation (Smith and Smith, 1975). Further, enough mercury had been released into the bay by 1959 that it would have taken a very long time for the bay to return to normality.

**FIGURE 1.3** *Takako Isayama and Her Mother*, ca. 1972 (W. Eugene Smith, Center for Creative Photography, University of Arizona: W. Eugene Smith Archive/Gift of Aileen M. Smith).

This recognition followed thirteen years of increasingly severe illness and debilitation of Yahei. Natsue was able to help him about their home but lost this ability soon after the 1972 images were captured (Smith and Smith, 1975). Since the Ikedas were initially very vocal in their skepticism of neighbors who claimed to be poisoning victims, their suffering extended outward into the community. "Not only did the poison corrode living cells, it insinuated itself into relationships between neighbors" (Smith and Smith, 1975).

Soon there were families in which several or all members were poisoning victims. The Smiths report that, as public awareness of the Minamata tragedy grew, homes of victims such as the Ikedas' became regular stops in "the parade of politicians and government environmentalists" (Smith and Smith, 1975). Interactions between the Minamata victims and the Chisso Company became so unsatisfactory that a Central Pollution Board was established in 1970 to decide issues, especially those pertaining to the lost victims (Figure 1.5). A few board decisions seemed to closely reflect the opinions of Chisso, and this further divided the community (Smith and Smith, 1975).

Some victims decided to pursue legal liability claims outside of the framework crafted by the board. A Fishermen's Union became involved with the confrontation. Families even brought their stricken members to the board meetings to remind board members of the real issue being confronted in Minamata. The stream of reporters and photographers into Minamata was constant and heavy.

Factions formed within the Minamata community. Reading the Smiths' Minamata book reveals labels such as the "old victims," "lost victims," "leave it up to others," and "make Minamata brighter" groups. The victims seemed to become secondary

**FIGURE 1.4** *Mr. and Mrs. Ikeda*, ca. 1972 (W. Eugene Smith, Center for Creative Photography, University of Arizona: W. Eugene Smith Archive/Gift of Aileen M. Smith).

**FIGURE 1.5** Members of the Central Pollution Board 1972 (W. Eugene Smith, Center for Creative Photography, University of Arizona: W. Eugene Smith Archive/Gift of Aileen M. Smith).

to causes and group opinions as disparate interests jockeyed for position and litigation dragged on. The head of one victimized family commented, "All this idealism about crusading 'for the world,' and so on. ... Hell, I took Chisso to court because this disease wrecked my family" (Smith and Smith, 1975).

As the Minamata situation drew international attention, the Smiths' focus remained steadfastly on individuals. The 1972 picture of Shinobu Sakamoto and her mother, taken as they waited for a train on their way to a United Nations meeting on the environment, exemplified the commitment to the personal tragedy (Figure 1.6). Shinobu's dress carries a protest message intended for the public; however, in Smith's photograph, this message is dwarfed by that conveyed in Shinobu's facial expression and posture—and her mother's handkerchief.

Shinobu was born in 1956, the same year her three-year old sister, Mayumi, died from mercury poisoning. Shinobu herself showed symptoms of *in utero* mercury poisoning within months of birth. Adding to her unresolvable difficulties and discomfort, she became to the press what the Smiths called "the maimed-but-struggling victim." Equally discomforting to victims such as Shinobu was the uncertainty about what would happen when the public's gaze inevitably shifted elsewhere to leave them alone with their handicaps (Smith and Smith, 1975).

> I went to Stockholm because I wanted everyone to know. Many people have never seen someone like me. ... I'm glad I went. Many people stared at me too, but I think the people understood a bit. My mother said so too. She told me so crying.
>
> **Shinobu Sakamoto (1972), Quoted in Smith and Smith (1975)**

**FIGURE 1.6** Shinobu and her mother preparing to leave the Minamata train station for the United Nations Environmental Conference in Stockholm, Sweden, 1972 (W. Eugene Smith, Center for Creative Photography, University of Arizona: W. Eugene Smith Archive/Gift of Aileen M. Smith).

## THE TRUTH ABOUT MINAMATA

Along with Rachel Carson's *Silent Spring*, the Smiths' Minamata images wrenched our collective attention away from the benefits of giddy industrialization, revealing the blatantly unacceptable human consequences of pollution. The Smiths ended their Minamata volume speculating:

> Historians might find in Minamata … —if humankind ever decides to assume true responsibility for its stewardship of this planet—that they are looking back into a kind

> of soul-force of courage, a force that might save our children from the plunders that began with the first industrial revolution.
>
> **Smith and Smith (1975)**

Such idealistic musings skirt the human truth of Minamata. The Smiths' images documented with subjective clarity the wrenched lives and wrecked families caused by mercury.

The Smiths' short marriage broke apart under the intense effort and stress associated with finishing the Minamata collection. Eugene Smith's health deteriorated, and three years after the volume's publication he died at the age of fifty-nine. He never witnessed the transformative impact of his images, nor did he live to understand his contribution to creating our present environmental ethos. He could not have known that this ethos would give rise to new fields of environmental science and extensive pollution legislation. However, unquestionably, he understood the human truth about Minamata.

## THE INTENT OF THIS SHORT BOOK

Hopefully this brief exposé about Eugene and Aileen Smith's Minamata photojournalism makes clear that our experience with mercury is much broader than our scientific knowledge of its chemistry and biological effects. It is the goal of this book to purposely step out of the conventional scientific framework for treating mercury pollution and to explore the wider human experience with mercury. As expressed in the opening quote for this chapter, to produce a technically rich book that only superficially mentions other features of our experience with mercury would result in an untruthful exploration of this important issue.

## REFERENCES


Bakir, Fatih, Salem F. Damluji, L. Amin-Zaki, M. Murtadha, A. Khalidi, N.Y. Al-Rawi, S. Tikriti, H.I. Dhahir, T.W. Clarkson, J.C. Smith, and R.A. Doherty. 1973. Methylmercury poisoning in Iraq. *Science,* 181:230–241.

Broad, William J. 1981. Sir Isaac Newton: Mad as a hatter? *Science,* 213:1341–1344.

Hughes, Jim 1989. *W. Eugene Smith: Shadow & Substance.* New York: McGraw-Hill Publishing.

Malm, Olaf 1998. Gold mining as a source of mercury exposure in the Brazilian Amazon. *Environmental Research, Section A,* 77:73–78.

Martinez-Cortizas, A., X. Pontevedra-Pombal, E. Garcia-Rodeja, J.C. Novoa-Munoz, and W. Shotyk. 1999. Mercury in a Spanish peat bog: Archive of climate change and atmospheric metal deposition. *Science,* 284:939–942.

Reardon, Ann Marie, and Hari K. Bhat. 2007. Methylmercury neurotoxicity: Role of oxidative stress. *Toxicological & Environmental Chemistry,* 89:535–54.

Smith, W. Eugene, and Aileen M. Smith. 1975. *Minamata.* New York: Holt, Rinehart and Winston.

Steichen, Edward. (2003). *The Family of Man.* New York: The Museum of Modern Art.

# 2 Dangerous Attractions
## *Mercury in Human History*

*Kris Lane*

### CONTENTS



Hg, from the Latin *hydrargyrum*/Greek *hydrargyros* = "liquid metal"

—*Oxford English Dictionary*

### INTRODUCTION

Demonized as an insidious toxicant today, mercury (element 80 on the periodic table) was not always feared and hated. Indeed, throughout most of human history this ambivalent, seemingly liminal substance and its compounds were highly prized. Archaeological remains and other sources make clear that mercury-based pigments, as well as medicinal preparations, were used for thousands of years across much of the globe. Elemental mercury figured, too, in several venerable forms of artistic decoration, such as gilding, and more secretively in alchemical experimentation, either to transmute base metals into silver and gold or to discover an elixir of eternal life.

Although mercury's natural attraction to gold, silver, copper, and tin was known for millennia, only after the discovery of massive silver deposits in what are today Mexico and Bolivia around 1550 did mercury's use expand to near-industrial scale. It was also after 1550 that mercury began to be shipped globally in large quantities for the purpose of amalgamation, a method of extracting precious metals from their ores by treatment with mercury to form an alloy.

Thanks to Spain's access to rich mercury deposits in southern Castile and highland Peru, as well as in Habsburg Slovenia, amalgamation was the standard refining practice throughout colonial and early national Spanish America. After the 1849 California gold rush and subsequent booms in western US, Australian, Canadian, and South African mining, California mercury, found near San José, became the solvent of choice until new processes sharply reduced its use around 1900. Even so, small-scale gold miners continue to practice mercury amalgamation throughout the world and into the twenty-first century, contributing as much as one-third of total global mercury pollution.

In addition to mining and refining, after 1850 mercury became a significant industrial ingredient, for example in explosives and mirror manufacture, as well as in synthetic fiber production. Mercury-based medicines and bleaching agents also sold well into the twentieth century, and silver–mercury amalgam became the standard means of filling tooth cavities with the rise of modern dentistry.

After World War II, and especially after widespread anti-nuclear protests in the 1970s and 1980s, coal-fired power plants became the major source of human-produced mercury in the atmosphere. Once released by combustion and carried by winds and ocean currents, trace amounts of mercury from coal began to concentrate in many of the world's remotest ecosystems, far away from mines or factories. Other anthropogenic (human-caused) sources of mercury pollution include garbage incineration and deforestation.

Although humans have long been aware of mercury's toxicity, only in recent times have scientists and policymakers made a conscious effort to evaluate mercury's spread on a global scale and begin to reduce its unwitting release. This reversal is quite recent in terms of human history and would seem a major awakening. Still, in an age of pressing and even compounding environmental concerns, keeping mercury pollution a top priority has been difficult despite well-publicized mass poisonings such as the one that occurred not long after World War II in and around Japan's Minamata Bay and another in Iraq in the early 1970s. In recent years growing levels of mercury in fish seem to be a stronger spur to action than past, localized disasters or documented poisonings of wildlife. European countries have taken the lead in banning industrial uses of mercury and also mercury exports, but whether mercury controls invite greater global cooperation than those surrounding climate change remains to be seen.

## MERCURY'S ANCIENT USES

Usually found on the earth's surface in its sulfide form as deep-red cinnabar crystals, elemental mercury was probably a rarer find for ancient humans than meteoric iron. Although scattered throughout the earth's crust, mercury sulfide only rarely occurs in high enough concentrations to be mined and refined. As an example of this rarity, much of the world's mercury now circulating in raw elemental form was produced by a single deposit in south–central Spain: Almadén in Don Quixote's fabled hills of La Mancha. Other concentrated sources have been found in Slovenia, China, Peru, and California, but none has proved as important or venerable as Almadén.

Prior to human involvement with mercury, nature had its own methods of redistribution or concentration. Most mercury emitted into the atmosphere and waterways came from volcanic and geothermal sources, plus surface erosion by wind, rain, frost, and other mechanical and chemical processes. It would be interesting to attempt to create a current world mercury budget to compare with a prehuman one, perhaps by searching core sediment samples for changes. Core sampling in remote lakes has already produced some surprising results for the ancient Americas. Mercury has always been an element related to human life, but to what degree have humans altered its dispersal across millennia, and to what effect?

### MERCURY-BASED PIGMENTS

Mercury sulfide, or cinnabar, is a dense, crystalline mineral with a bright red color that in the Western world came to be known as vermillion. Vermillion oil paints made from cinnabar were in common use until the late twentieth century, and in the eighteenth and nineteenth centuries British firms such as the Hudson's Bay and Northwest Companies traded vermillion face paint to the many native peoples of North America for beaver pelts, deerskins, and other forest products. This face paint may sound dangerous, but in the form of vermillion ointment, mercury was not especially toxic unless it entered open wounds or was burned, releasing dangerous fumes. Much more troubling was European hat makers' use of mercury salts to treat the beaver pelts produced by this trade. Some hatters, such as those in Lewis Carroll's hometown, allegedly went mad after vaporizing mercury in the felting process.

Even in its sulfide form, mercury readily attaches itself to gold, silver, copper, and tin. It forms a chemical bond and acts as a solvent. Given these facts, it is understandable that powdered cinnabar and vermillion paste were used to paint objects made of these metals. Indeed, such objects have been found in archaeological sites from ancient China to the Andes. In Peru, Ecuador, and Bolivia many gold, silver, and copper-alloy masks and similar artifacts dating to the first centuries of the Common Era show traces of cinnabar.

In pre-Columbian times most South American cinnabar likely originated in the mountains east of Lima at a site called Huancavelica, later exploited by the Spanish (See Figure 2.1). Cinnabar from lesser deposits in Mexico, Guatemala, and elsewhere was also traded widely, but only in places where urban cultures left substantial traces has it been routinely found. Artifacts in Mexico bearing traces of cinnabar suggest that the pigment was well known far beyond its source regions, but no source was

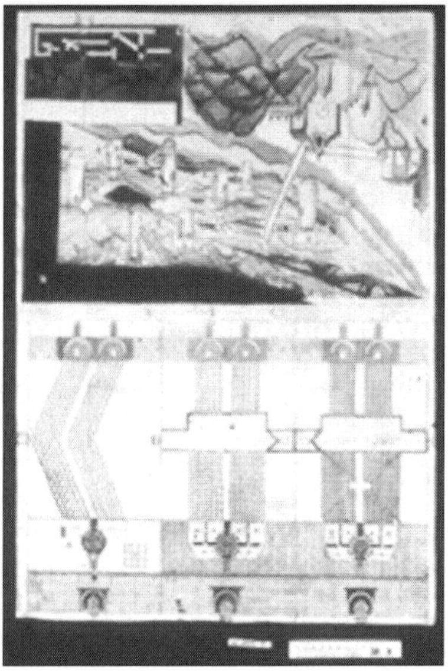

**FIGURE 2.1** The mercury mines and refineries of Huancavelica, Peru. In this eighteenth-century schematic, mine shafts seem to connect to churches in a topsy-turvy landscape. Below are the characteristic earthenware condensing jars used to recover liquid mercury from heated cinnabar. Courtesy of the Archivo General de Indias, Seville, Spain.

as concentrated as that of Huancavelica in Peru. Mexico's cinnabar deposits were thoroughly explored in colonial times, but never proved economical for mercury production.

Similar cinnabar traces have been found on ancient artifacts from the Mediterranean, Europe, North Africa, and the Middle East, mostly derived from Spanish and perhaps also Slovenian cinnabar. By itself or mixed with oils and other substances, cinnabar was used to decorate walls, ceramics, textiles, and the human body. As a chemically stable sulfide pigment, mercury's toxicity was relatively minor.

## Mercury-Based Medicines

Until recent times, most humans believed that sickness was caused by either the presence or absence of poison in the body or an imbalance in a sort of fluid budget. Acts of witchcraft were described in these terms in many world cultures—victims were poisoned, or made to "dry up." A wide range of therapies were thus developed with the aim of detoxifying the patient or restoring the balance of "humors," which in the Western tradition included blood, black bile, choler (yellow bile), and phlegm. Balance was also figured in terms of hot/cold, dry/wet, and other paired extremes that could vary according to age, gender, and personal temperament—partly determined by one's astrological sign. The physician's or healer's role was to diagnose types of poisoning and specific imbalances, then to prescribe and apply the proper medicaments to detoxify and restore the individual patient's normal "humoral" balance.

We do not know about mercury in medicine in the ancient Americas or sub-Saharan Africa, but therapies throughout Eurasia tended to use mercury in two ways: externally, usually as a caustic sublimate to treat wounds and infections, and internally, as a means of encouraging salivation to restore humoral balance. In Taoist, and to a greater degree Ayurvedic, medicine that prevailed in India and parts of East and Southeast Asia, mercury use became overtly mystical, associated (wrongly one must presume) with immortality. In some Ayurvedic texts, mercury is associated with the highest ranks of godhead and serves as a kind of elixir. We may think this absurd, but mercury's power derived in large part from its dangers.

How did ancient healers obtain liquid mercury? As nature would have it, mercury is released from cinnabar at relatively low temperatures (c. 360°C), and thus it was not long before curious artisans or tinkerers learned to trap and condense its vapors and produce pure quicksilver. Inhaled mercury vapor is highly toxic, and presumably some early experimenters discovered this the hard way, losing their teeth, salivating heavily, and suffering temporary paralysis and blindness, if not death. Liquid mercury is far less toxic but is easily blended with salts to form caustic as well as violently poisonous compounds. Some of these compounds, such as the paste called "Soliman," or "Turkish ointment," served as astringents and antiseptics and were carried in soldiers' medical kits from before the era of the Crusades. After Columbus's voyages, Turkish ointment was also used to treat syphilis sores. At the other end of the spectrum "corrosive sublimate," or mercury chloride, was widely known to be a deadly poison.

As these several health-related uses of mercury derived in part from alchemical experimentation and the search for elixirs of youth, we shall turn next to this, mercury's most mysterious historical chapter. It should be kept in mind that mercury-based medicines were widely used until the mid-twentieth century, and some quite effective ones are still being developed. In this sense, mercury's twin abilities to poison and heal render it the archetypal pharmakon—a most ambivalent and, by some accounts, still magical substance.

## THE AGE OF ALCHEMY

As far as we know, most premodern humans believed that metals were some kind of divine residue. In the ancient Andes, gold was described as "sweat of the sun" and silver as "tears of the moon." Many cultures also believed that transforming metals from their raw (native) or ore forms into workable materials was a sacred task, to be performed by chosen individuals or clans. Almost worldwide, meteoric iron, as well as native gold, silver, or copper, figured in both cosmology and material culture. (Historians later assigned "metallic" ages to various civilizations, e.g., bronze, iron, gold, gilded, etc.)

In general, as the late historian of religions Mircea Eliade and others have noted, the esoteric skills required in processing and manipulating metals were not only assigned spiritual significance but also were often carried out in great secrecy. In some places, notably parts of sub-Saharan Africa, metalworking was monopolized by powerful clans who tended also to control trade networks for metal tools or currency. Similar mystery surrounded metallurgy in other cultures, most notably in Eurasia. Here mercury, since it was capable of transforming or at least combining with the precious metals, was quickly drawn into the secretive metalworker's kit.

## Ancient Alchemy

As it is liquid at room temperature, mercury was not widely considered a metal, at least in ancient Eurasia, but rather an ambivalent substance possessing select metallic properties, such as density and brilliant, reflective color. Mercury's affinity with gold, copper, silver, and tin led some experimenters to believe it was an ingredient in these metals' constitution. In Western alchemical thinking—which contrary to popular belief was not simply an obsessive, futile quest to transmute base metals into noble ones—mercury and sulfur, which also displayed ready affinity for all but gold, were the constituent "elements" of all metals; each metal was a discrete combination of the two as rendered intractable to humans by fickle nature or patron deities. The actions of various acids on the metals were also well known from an early date, and the philosopher's stone was part formula for transmutation (key to wealth and power), part formula for the elixir of life (key to immortality). Alchemy's philosophical underpinnings could be said to match well with mercury's own "chemotherapeutic" ambivalence: what cures also kills.

Alchemical treatises were produced in many ancient cultures in Eurasia, from China to the circum-Mediterranean, and in every case mercury was a key component, usually part of the *tria prima* with sulfur and salt. Judged as inscrutable by many later chemists, these treatises often resorted to esoteric language and a wide range of bizarre symbols, including hermaphrodites, "squared circles," and so on. Alchemical symbolism related to religious or mystical concepts had nothing directly to do with the chemical properties of mercury or other metals, and thus won the opprobrium of both orthodox religious and enlightenment critics. Others disagreed. In the twentieth century, the psychologist Carl Jung, an early disciple of Freud, considered classical alchemical writings as a kind of key to the human mind and social relations.

## Early Modern Alchemy and the Rise of Chemistry

In the traditional Scientific Revolution narrative (admittedly much revised nowadays), modern chemistry arose like a phoenix from alchemy's ashes. Beginning in the sixteenth century, alchemy came to be associated with greed and ignorance, denounced as sinful by preachers and stupid by practical renaissance men such as the German metallurgist Georgius Agricola. Especially in Protestant Europe, a movement was afoot in favor of direct observation and scrupulous experimentation. Received wisdom, however old, ought to be tested, even overturned, by employing improved technologies in metallurgy and optics, freely exchanging information in print, and unabashedly seeking to unlock nature's secrets. Alchemy's secrecy and mystification flew in the face of this trend and even in the Catholic world, princes who clung to alchemical notions suffered censure by the Inquisition. The scientific method, which required that an experiment be repeatable by anyone following the proper recipe, seemed poised to spell alchemy's doom.

Interestingly, many seventeenth-century metallurgists, including the priest Alvaro Alonso Barba of Potosí, and even Sir Isaac Newton, who was briefly employed in England's royal mint, tried to follow a middle path. They believed that the massive corpus of alchemical texts bequeathed to them by posterity could not all be wrong. Experimentation and observation were required to make advances, as both admitted freely, and received wisdom had to be tested, but much was to be learned from the alchemists' millennia of tinkering.

Barba's work in Potosí, a famous silver-mining town in what is today Bolivia, reveals a deep fascination with mercury, with which he had unusually copious experience, but also a considerable practical knowledge of its properties under many conditions and in many combinations. Barba cited key alchemical authorities such as Raymond Llull in his c.1627 treatise *The Art of Metals*, and as a result saw his work suppressed by the Inquisition. In faraway North America, the Harvard-educated metallurgist George Starkey, one of Barba's contemporaries (although Barba did not apparently know Starkey's work), remained by contrast "stuck in the past." His widely read works were far less practical in tone and bore few traces of the emerging scientific skepticism made popular by Francis Bacon. Even Newton, working at the end of the seventeenth century, lacked much of the practical knowledge and willingness to assert new ideas typical of the Catholic priest and inveterate "lab rat" Barba. Newton was heavily influenced by Starkey.

By the early eighteenth century, European tinkerers and scientists began to notice that more than the seven sacred metals (each associated with a celestial body and divinity) were in existence. Bismuth was an early challenger, but since it was difficult to refine, it was only with the discovery of South American platinum around 1750 that major experimentation with metals led to the breakthrough in taxonomy—and understanding—required for the development of modern chemistry. The new taxonomy at last identified mercury as but one of many metals that, despite natural combinations with other substances, could be reduced to discrete, "immutable" elements. Alchemical claims of a metallic continuum, or spectrum, were finished when in the 1780s Lavoisier used Priestly's newly discovered element of oxygen to create a torch powerful enough to fuse platinum. Suddenly it seemed as if all metals, mercury being only one of many, had been tamed by the handiwork and ingenuity of determined and intelligent mortals. Mercury was thoroughly demystified; humans were finally in control. Or so it seemed. Worldwide, mercury continued to be mined and applied to medicine and industry, most importantly the refining of Spanish American silver.

## MINING MERCURY

Mercury's occurrence in the earth's crust is about 0.07–0.08 parts per million, close to that of silver. Like silver, mercury is rare enough that it can only be profitably mined in concentrated deposits, usually massive sulfides. Ancient deposits of cinnabar were exploited in many parts of the world, but significant concentrations were limited. Mercury mining, or rather the mining and processing of cinnabar, is best documented for Almadén, Spain (see Figure 2.2), although equally ancient deposits in China were likely exploited on a considerable scale.

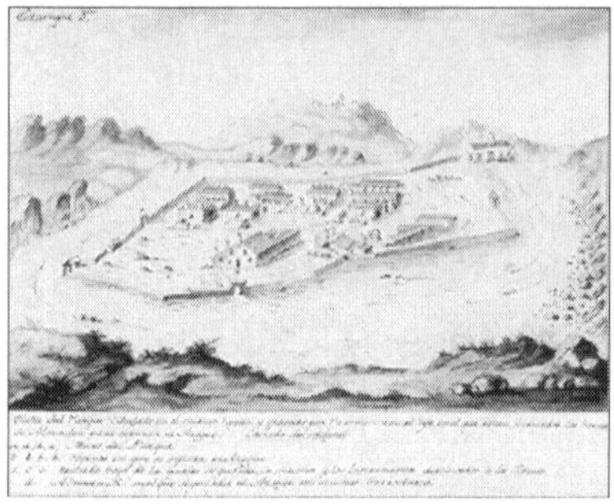

**FIGURE 2.2** *Mercury as Royal Treasure.* This late eighteenth-century watercolor depicts the new refinery complex at Almadén, Spain, including the building at front (d), the "Royal Storehouse in which the mercury is safeguarded with many locks." Courtesy of the Archivo General de Indias, Seville, Spain.

The mercury mines of North America seem to have been only lightly exploited before Europeans arrived, amounting to little more than surface collection of cinnabar. Recent studies of lake deposits near Huancavelica, Peru, suggest substantial cinnabar mining in pre-Columbian times, but the extent of liquid mercury production is unknown. Spanish mercury production, by contrast, is well documented. It rose rapidly in Roman times, when mining was a major occupation throughout the Iberian Peninsula. Despite the presence of considerable gold and silver deposits in Spain, however, we have no evidence of Roman employment of large-scale mercury amalgamation in recovering these metals. Recovering gold dust by gravitational methods with water was instead common in the north, at Las Médulas, and recovering silver and copper by smelting was common in the south, at Rio Tinto and other sites.

## Mining Hazards

Mining mercury, or rather cinnabar, presents a range of health hazards. First, it shares with other types of mining the possibilities of death or injury from falls, cave-ins, foul air, explosions, flooding, and other objective dangers. Worse, according to historian Kendall Brown and others who have examined manuscript records, was the effect of breathing cinnabar dust. Since mercury's bond with sulfur is weak, chipping or blasting cinnabar yields a mix of finely vaporized mercury and sharp crystal shards, both of which hang in the air in the limited spaces typical of underground mines. Constant inhalation of these mercury-rich particles usually led to an ailment combining characteristics of emphysema and partial paralysis. Acute mercury poisoning from mine work caused hair and tooth loss, reduced vision, insomnia, heavy salivation, and tremors. Due to these extra hazards, few miners volunteered to extract or process cinnabar, and most premodern mines and refineries were staffed by coerced workers, including slaves and convicts. In Spain convicts faced the galleys, service in the sun-baked fortresses of North Africa, or the mercury mines of Almadén (see Figure 2.3).

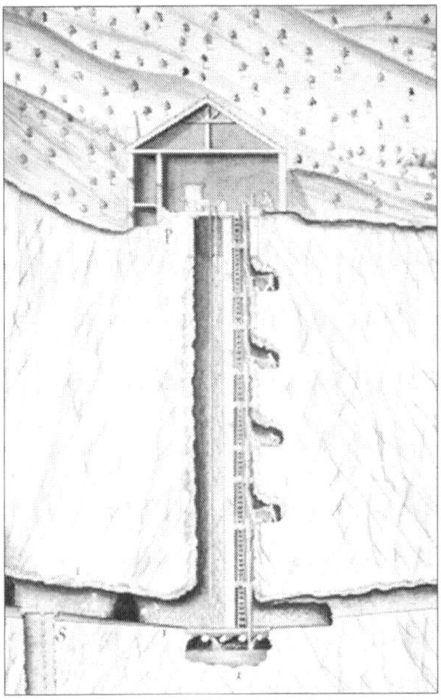

**FIGURE 2.3** *A World Going On Underground.* This late-eighteenth-century image showcases Spain's Almadén mercury mines' newly improved shaft and adit system, complete with integrated, human-powered drainage pumps. Most workers were condemned convicts. Courtesy of the Archivo General de Indias, Seville, Spain.

## Refining Hazards

Refining mercury from cinnabar was perhaps equally hazardous, as it entailed heating the excavated mineral in furnaces and reducing the vaporized mercury to liquid by condensation. Even using the very best modern equipment, containing mercury vapor is difficult, but in early modern times, the use of poorly sealed stone ovens and ceramic condensing vessels meant that substantial mercury was released into open air where it was no doubt inhaled by refinery workers and neighbors. Apparently cognizant of this problem, the owners and contractors responsible for Spain's mercury mines lived at considerable distance from the refineries.

Once refined, mercury from Spain, Peru, and Slovenia was packaged in flasks and skins, then crated for carriage by cart, mule, or even llama. Mercury was valuable enough to be monopolized by the Spanish crown beginning in the sixteenth century, partly in hopes of keeping tabs on silver refiners by calculating consumption. Spain proved too profligate with its reserves, however, and magistrates periodically auctioned the exclusive rights to mine and distribute Almadén's and then Huancavelica's bounty to bidders, most famously the Fugger family of Augsburg in southern Germany. Ships carrying mercury regularly crossed the Atlantic, mostly to supply the mines of Mexico, but also those of Honduras and New Granada (modern Colombia), and some of these vessels were wrecked in the Caribbean, leaving behind still-visible pools of mercury among the coral reefs.

## AN AMERICAN AMALGAMATION REVOLUTION

The year 1554 marked a major turning point in the global history of mercury. In this year, a merchant from Seville, Bartolomé de Medina, developed a process for amalgamating silver ores in Pachuca, Mexico, a dusty camp not far from Mexico City. Medina's method, eventually expanded into what became known as the "patio process," was adopted throughout Spanish America and used until the nineteenth century. It allowed for finely crushed silver ores to be refined over the course of many days by the timed and measured addition of salts and mercury. Amalgamation itself was not new, but this innovative series of steps was. The mix was stirred by indigenous workers and mules, aided by solar energy, and eventually the resulting amalgam was separated from waste mineral to be washed and fired to yield nearly pure silver "sponges," which were then refired and poured into bar molds. Medina's innovation was especially important in the mining districts of Mexico and the Andes in that it required far less fuel than smelting, which sped deforestation. Again, mercury's use had two sides: it saved trees while yielding silver, yet it also poisoned workers and animals and fouled air and streams.

### Mercury in Spanish America's First Great Mining Cycle

Medina's discovery quickly transformed Mexico into a significant world producer of silver, but it was distant Potosí in what is today highland Bolivia that most benefited in the shorter term. After about 1570, the recently discovered mercury of Huancavelica, Peru, was applied to the rich silver ores of Potosí in what officials described as a God-given "marriage." By the 1590s amalgamation yielded tens of millions of ounces of silver annually, and by 1610 Potosí was home to well over 100,000 people, making it one of the Western world's largest cities. Given the fact that mercury vapor was being pumped into the air almost constantly, along with lead, zinc, and many other toxic substances, Potosí was likely among the world's most polluted cities as well. Today historical archaeologists have resisted the temptation to dig sites in the old colonial city thanks to visible layers of mercury and other metals in the soil. It was here in Potosí that the priest Alvaro Alonso Barba tried numerous recipes aimed at

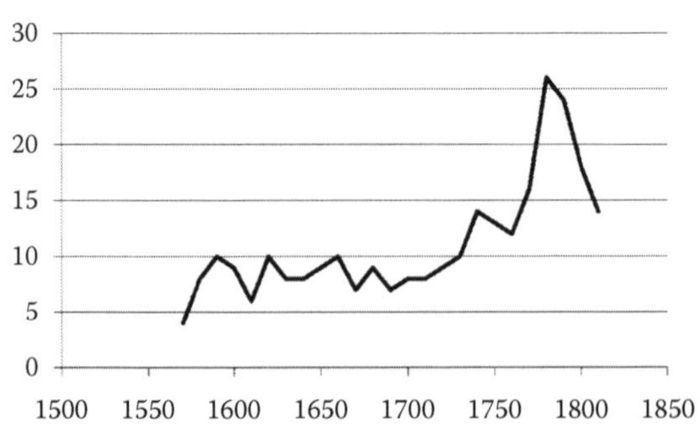


**FIGURE 2.4** Production figures in thousands of quintals, for Almadén, Idria, and Huancavelica (1570–1810).

recovering more silver and losing less mercury, all recorded in his learned treatise, *The Art of Metals*.

Once Potosí's deposits began to play out in the early seventeenth century, Spanish attention shifted back to Mexico, where the old silver districts of Pachuca, Taxco, Zacatecas, and Guanajuato were revived thanks to new investment and improved infrastructure. New mines were also opened at San Luis Potosí, Real del Monte, Parral, and Chihuahua. Critical to all of these mining ventures was the mercury of Spain's mines of Almadén, shipped annually across the Atlantic and carried overland by mule. In Mexico the "patio process" was expanded in scale and adapted to different types of ores. Tens of millions of coins minted from Mexican silver soon became the basis for many world currencies, including the US dollar and Chinese yuan.

## Mercury in Spanish America's Late Colonial Revival

Due to population decline and played-out deposits, Spanish America's mining industry faced many difficulties in the late seventeenth and early eighteenth centuries. Crown supplies of mercury from Almadén, Idria, and Huancavelica all sputtered, and many ships were lost to pirates and privateers. In some places miners and refiners took to smelting. What turned around this decline in the mining industry was a series of reforms initiated by Spain's new Bourbon monarchs and their ministers, all of them inspired by the Enlightenment and Louis XIV's notions of absolutism. Spain's centralizing kings devoted much attention to expanding and rationalizing the mining industry as a means of meeting ballooning military expenditures. Mercury, a crown monopoly, was key.

The result of mining and refining reforms established by the Bourbons, beginning with Philip V (1700–1746) and continuing through the reign of Charles IV (1788–1808), was renewed output of mercury, especially at Almadén, but more importantly a huge revival of the silver mining industry, from northernmost Mexico to Chile. By the time Napoleon's forces invaded Spain and removed Charles IV from power, Mexico alone was minting nearly 20 million "silver dollars" a year.

Its price reduced, thanks to crown intervention, mercury was consumed at a higher rate than ever. However, a literal cloud of mercury vapor settled over the famed silver cities of Spanish America. Thus, in generating a huge portion of the world's money supply, generations of poor workers and their families, most of them Native Americans, absorbed this unwitting "cost" of production, and many generations of enslaved Africans suffered from work in the refineries and mints.

## Mercury in the California Gold Rush and Western US Bonanzas

Just as California was being incorporated into US territory following the invasion of Mexico in 1846, major cinnabar deposits were found just south of San José. The mines were called New Almadén, and a later find was called New Idria. The finds were soon overshadowed by the historic discovery of gold not far away at Sutter's Mill near Sacramento, but the discovery of mercury in California, so close to major precious metals deposits, was to prove extraordinarily important for the development of the mining industry in the western United States. Following a modified and greatly expanded version of the Spanish "patio process," North American miners and refiners soon applied New Almadén and New Idria's mercury to the massive silver ore deposits of the Comstock on the California–Nevada border, then to the many silver ores of Colorado, Idaho, New Mexico, and other mountain states.

California mercury production rose so quickly that it overshadowed production in Spain and Slovenia, both of which in the mid-nineteenth century were monopolized by the Rothschild family. The Huancavelica deposits in Peru were in such decline that mercury from California was exported to aid silver refiners in Peru and Bolivia as well as the much nearer mines of Mexico. Conditions in New Almadén were never so bad as at Huancavelica under the Spanish, but mineworkers and refiners, many of them Mexican–Americans, suffered the characteristic symptoms of mercury poisoning consistent with their trade. As always, cinnabar dust in the mines and mercury vapor in the refineries were the main culprits.

Around the turn of the twentieth century, mercury stopped being a major refining ingredient in precious metals production. Chemists and other scientists had at last developed industrial-scale alternatives, the most important of them using sodium cyanide to dissolve gold. Cinnabar deposits were one by one abandoned worldwide by the 1990s (with the possible exception of China). As of this writing, considerable mercury is produced as a byproduct of large-scale gold mining, which has expanded tremendously in tandem with the global economic crisis. In an ironic twist, given the history of mercury just recounted, modern industrial gold processing requires no mercury, but instead yields it.

The story does not end there: Since the 1930s, small-scale gold miners have been continuing to amalgamate gold (and to a lesser degree silver) ores, fouling air and streams and poisoning many people along the way. They have not done so on purpose, but with every gold price spike, as occurred in the late 1970s and late 2000s, mercury consumption by small-scale gold miners, particularly in developing countries but even in the United States and Canada, has spiked with it. In another irony many of these miners buy their mercury from wholesalers who purchase it from first-world recyclers of batteries and other industrial junk.

## INDUSTRY'S CALL

With the rise of modern chemistry a great variety of new uses for mercury were discovered. Dense and not liable to evaporation, freezing, or corrosion under ordinary circumstances, mercury proved ideal for making thermometers and barometers. Mercury is also a conductor and was soon adapted for use in switches, particularly thermostats. A major breakthrough occurred in the 1890s when scientists discovered that mercury vapor could produce significant light when sealed in a vacuum tube and charged. Industrial production of such vapor-filled tubes began in the late 1930s and as a result fluorescent lights, far more efficient than their incandescent ancestors, became standard after World War II. Fluorescent household lighting was generally resisted for aesthetic reasons until a new generation of bulbs, popularized by huge chain stores, became available around the year 2000. Some environmental activists have hailed fluorescent lighting for its relative efficiency, but little attention has been paid to proper disposal of bulbs. In industry, as in medicine and mining, mercury's benefits and drawbacks must be placed in the balance.

## Scientific Instruments

The first major scientific instrument to employ mercury was the barometer, used to measure atmospheric pressure. The first mercury barometer was developed in 1643 by a student of Galileo, the physicist Evangelista Torricelli. Others had attempted to measure atmospheric pressure by creating columns of various fluids, including water, but mercury proved most suitable due to its high density. A relatively compact, 33-inch column could be supported in a mercury reservoir by normal atmospheric conditions. The base unit of "one atmosphere" was set at 29.9 inches of mercury, and this system is still used in the United States instead of the now world-standard metric, expressed in millimeters of mercury. Early experiments with barometers showed their value also as altimeters, but since they were unwieldy they were rarely carried except on major scientific expeditions.

Blown-glass mercury thermometers were to prove far more portable than barometers, and hence potentially more dangerous. Historians of science credit German physicist Daniel Gabriel Farenheit with developing the blown-glass mercury thermometer sometime in the 1710s while living in the Netherlands. In 1724 Farenheit published a description of his namesake temperature gauge, still used in the US. Swedish astronomer Anders Celsius offered an alternative scale in 1742. The clinical, or "fever," thermometer used to measure body temperature was developed in England in 1866 by physician Sir Thomas Clifford Allbutt and remained standard until quite recently, when digital thermometers became economical. After 2007 many European countries banned production and sale of blown-glass mercury thermometers, although other countries still produce them and millions remain in use. In lieu of alternatives, it is difficult to imagine how else but by mercury thermometers body temperatures might have been efficiently measured over the last century and a half. As in so many other instances, mercury proved dangerously useful, if not absolutely necessary.

## Lighting, Switches, and Batteries

In addition to the fluorescent lightbulb, inventors and industrialists employed mercury in making switches, triggers, and batteries. Mercury switches were until recently used to sense off-kilter movements, including the tilting of pinball and vending machines, but more commonly in the manufacture of thermostats. Mercury was used in addition to bimetallic filaments to ease the transitions between the "on" and "off" states of heaters and air conditioners. Many buildings still use these reliable devices, although like thermometers they are being phased out with digital replacements. The same tilt feature of mercury switches has been used to make triggers for various kinds of bombs, including landmines, but this nefarious application has been mostly limited to improvised devices.

By contrast, fulminate of mercury [$Hg(CNO_2)$], a compound first identified in 1800, has had a long and violent history. Beginning in the mid-nineteenth century, it was applied to industrial arms manufacture. Small amounts of mercury fulminate were enclosed in caps and built into bullets, rendering flints and flashpans obsolete. Until lead-based alternatives became available in recent years, other uses of mercury fulminate included blasting caps for dynamite, the celebrated (and despised) invention of Alfred Nobel. The amount of mercury used in bullet and blasting cap manufacture was relatively negligible, and probably far less dangerous than the blasts it helped create, but weapons stockpiles and old manufacturing facilities continue to present a notable mercury hazard.

Mercuric oxide, usually used in combination with other metallic compounds, was employed beginning in World War II in battery manufacture. Although not rechargeable, mercury-based batteries proved more reliable than any alternatives, including the alkaline cells that have largely replaced them. Mercury-based batteries, like mercury thermometers, are being banned worldwide, mostly due to disposal hazards rather than leakage or any other direct toxic effect on users. In a world increasingly reliant on portable electronics, the search for alternatives to mercury-based batteries is one of several areas of intense research.

In all of these industrial applications, mercury has proved versatile and dangerous in roughly equal measure. Only with growing consciousness of mercury's lingering hazards were inventors and industrialists spurred to develop alternatives.

## From Industrial Handmaiden to Angel of Death

From 1850 to 1950 mercury's many applications seemed to promise unlimited progress with the advance of industrial societies and intensified interest in research and development. The saying "better living through chemistry" was as applicable to mercury as it was to any other element. Unfortunately, mercury's other face, its poisonous aspect, began to reappear, especially after World War II. Mercury sulfate and other such compounds proved to be handy catalysts in the manufacture of polymers, fertilizers, and basic chemicals, many of them developed during wartime. In postwar Japan, industrial production soared as the country rebuilt and reinvented itself as a peaceful powerhouse. The Chisso Company was a pioneer in the manufacture of fertilizer and other industrial chemicals, and ballooning demand for its products combined with negligence and ignorance of waste handling led to the so-called Minamata disaster, first documented in 1956.

Minamata Disease, as it became known, was a debilitating ailment caused by the unwitting consumption of methylmercury that had accumulated in fish and shellfish in Minamata Bay. Methylmercury, a highly toxic organic compound discussed at length in other chapters of this volume, had formed as a by-product when mercury salts such as mercury sulfate were used as catalysts to manufacture acetaldehyde, a chemical used for making vinyl, acetates, and other substances. Minamata Disease, which killed as many as 2,000 people and severely affected the lives of many more, awoke the world to the dangers of mercury poisoning. Continued problems with mercury poisoning in and around Minamata Bay recurred into the 1960s and led local citizens to demand government oversight of heavy industry. Despite the global awareness of mercury poisoning that resulted from the Minamata Disaster, documented with harrowing immediacy by photographers W. Eugene Smith and Aileen M. Smith, ignorance of mercury's ubiquity continued to prevail.

## THE PRICE OF POWER

Given the central and obvious importance of elemental mercury in the gold and silver refining industries described at length above, it may come as some surprise that by far the most significant polluter of recent times—with the exception of some massive volcanic episodes—has nothing to do with refineries or industrial uses of the liquid metal itself. Coal-fired power stations, the number of which has grown exponentially since the antinuclear power movement of the 1970s and 1980s, have become the world's most prolific sources of mercury contamination. Noting this shift to coal with its mercury pollution is not meant to suggest that the antinuclear movement was misguided, as several "meltdowns" and near misses proved otherwise, but simply that demand for electricity did not diminish, and in fact increased after the late twentieth-century energy crisis. In countries such as the United States and China, which boasted massive reserves, coal emerged as the major source of kilowatts.

As nature would have it, coal contains trace amounts of mercury, not enough to cause a stir when small amounts are burned, but enough to pollute oceans and continents when consumed in the quantities needed to power nations of millions and even billions. Some attempts to control mercury emissions have been made at the national level (including by the US Environmental Protection Agency and US Department of Energy), and preliminary discussions have begun internationally. Still, removing mercury vapor from power plant exhaust is difficult, and coal types containing no mercury have yet to be found. Industry advocates have thus stressed emissions reduction rather than elimination. Either way, the scale of the problem has only exploded in recent decades. As in so many other cases described above, human demand for things of beauty, health and hygiene, wealth, efficiency, and now electrical power has intersected with mercury, a metal that has served as both a blessing and a curse.

## THE AGE OF ACTIVISM

Environmentalism as a political movement is a recent phenomenon. Human concern for health, by contrast, is at least as old as Austria's prehistoric Ice Man, who carried medicinal herbs and talismans to his death. As we have seen, mercury has for most of human history been associated with health and progress despite knowledge of its poisonous side effects. Only recently have citizens of many parts of the world pressured not only local and national governments but also international oversight bodies to ban many forms of mercury production and dissemination. To reduce its toxic effects on the environment has required a near demonization of mercury, a reversal of its historic role in human affairs. Whereas it was once understood that mercury, like alcohol or many other substances, could kill as well as heal, today it is hard to find anyone willing to defend any of mercury's "good" qualities. Such may be the nature of activism, but activism, too, may be scrutinized under the historian's lens.

## Premodern Concerns Regarding Mercury Hazards

Ancient writers were aware of mercury toxicity, although they presumably did not understand how exactly it poisoned the human body. As with other pharmakoi it was dosage and form that mattered, not the base material itself. Again, alcohol and other such substances were understood to be toxic in excess but potentially beneficial in moderation or when administered to address a particular ill. Most arguments among physicians up to modern times centered on how mercury was prescribed or handled, not whether it should be prescribed or handled at all.

Only with the expansion of the Huancavelica mercury mines in highland Peru in the 1560s do we find an analogue of modern antimercury activism. As historian Kendall Brown has shown, a number of Catholic priests charged with administering the sacraments to Native Americans drafted to work in the mines began to denounce mercury mine and refinery work for its atrocious health effects. They tended to emphasize cave-ins and other hazards as well, but they hoped to win the king's attention by describing symptoms of acute mercury poisoning that would have been familiar to many in Spain. Worried that their work in saving souls was endangered by Spanish greed (mercury being used for silver amalgamation at Potosí and elsewhere) these writers petitioned the Spanish crown to end the labor draft. As it happened, the proposed alternatives, replacing native Andean workers with either criminal exiles or enslaved Africans, did not prove attractive.

The only thing that lessened indigenous suffering at Huancavelica's mercury mines was the playing out of rich deposits. What about the massive gold and silver refining operations described above? Did they draw similar denunciations? Interestingly, daily use of mercury in silver and gold amalgamation throughout the Americas up to about 1900 was virtually never singled out for reform, despite its obvious health effects. Why? As in the case of Minamata's Chisso Company, those who stood to profit usually chose to ignore or suppress evidence of mercury's toxicity. Uninterested in protecting racially or socially marginalized subjects and citizens with such wealth at stake, governments tended to tow the industry line.

## Modern Activism vs. Mercury Pollution

The pitch and scope of global antimercury activism has only recently expanded, despite the now fifty-plus-year-old lessons of Minamata. Unlike recent antideforestation and climate change campaigns, however, most antimercury activism has derived from everyday fears about personal health rather than more abstract concerns over the global environment per se. Many antimercury activists in the developed world now focus on mercury in fish, dental amalgam, and vaccine preservatives. Others target coal-fired power plant emissions, small-scale gold mining pollution, disposal of mercury-based batteries, and many other concerns. From a historian's perspective, human feelings about mercury have reached an all-time low; the liquid metal that once fascinated and healed now seems only to be a deadly toxicant that has sneakily infiltrated everything.

Environmental activism, or the preservation of natural habitats against, for example, logging, mining, hunting, farming, snowmobiling, and other transformative or simply disruptive human activities, is only about a century old in the United States. The idea of parks as national patrimony grew out of older European royal hunting preserves. The notion of setting aside international parks or ecological preserves is an even more recent phenomenon, at least as regards fragile regions in the so-called developing world. Some such preserves in South America, Africa, and Southeast Asia, several of which are vulnerable to poor gold-seekers using mercury, are now traded as if they were marketable products.

Despite flickers of concern in the sixteenth and seventeenth centuries, modern antimercury activism is perhaps best understood in relation to the Industrial Revolution, particularly in its later phases. Early nineteenth-century industrialization was sharply criticized by anarchists, socialists, communists, and many others, but usually for its social and economic rather than strictly environmental consequences. Still, many Romantics were appalled by what they regarded as industry's invasive scale and sterility, plus its allegedly deadening effects on the human imagination. Even in the nineteenth century writers and artists were keen to point out the relationship between industrialization and pollution, yet none focused on mercury or other heavy-metal contaminants.

Only with the explosion of the petrochemical industry after about 1910 did industrial pollution come to be noticed, but this concern about pollution was to a large degree interrupted by the First World War, the Great Depression, and finally World War II. Only the Roaring Twenties, in which many nations raced to expand production based on new fossil fuels, gave a clue as to what might happen later. Interestingly, coal was in many places seen as a fuel of the past rather than the future. After the devastation wrought by World War II, US, European, and Japanese industries expanded or rebuilt, and amid rapid postwar growth the scale of industrial pollution soon dwarfed anything that had come before.

Worker safety rules and regulation of effluents and exhaust lagged far behind the intense pace of industrial expansion, and the corporate rhetoric of the time (for example, with cigarettes and leaded gasoline) was relentlessly upbeat. Having survived the Depression and two world wars, many people were anxious to believe that their lives would indeed be made better through chemistry. Aided by advertising, medical advances, not to mention full employment, comfortable housing, and access to an unprecedented range of consumer goods, seemed to mostly bolster this optimistic view of industrial modernity. Who was to complain if batteries contained mercury when they made music, lighting, and other things portable? And was it not better to have cavities filled with amalgam than to lose one's photogenic smile?

Thus it came as something of a shock that the first world's growing reliance on industrial chemicals was, in folk language, "akin to holding a bear by the tail." The Minamata disaster of the mid-1950s threw industrial negligence and government complicity into sharp relief, prompting not only the formation of activist groups seeking redress for specific spills and other incidents but also generally raising popular consciousness as to the potential dark side of industrial development. Mercury was briefly in the spotlight, and especially after the late 1960s people living in industrialized regions worldwide began to notice that living next to smelters, factories, power plants, and so on was probably not good for their or their family's health. Yet the fact that industrial development provided jobs left many ambivalent, uncertain of what to do.

Although there had been outspoken individuals since the time of wilderness activist John Muir (1838–1914), it took flaming rivers and the industrial-scale chemical horrors of Vietnam to raise the environmental consciousness of many North Americans. Episodes abroad, such as the inadvertent poisoning of thousands of starving Iraqis who ate mercury-treated seeds sent as food aid, only heightened popular distrust of industry and government, what Dwight D. Eisenhower had famously called the "military–industrial complex." Out of the inevitable backlash came the formation of the US Environmental Protection Agency, along with many nongovernmental organizations such as Greenpeace and the World Wildlife Fund. On the other side of the Iron Curtain, meanwhile, criticizing industrial pollution was not tolerated (although growing popular awareness of it helped erode support for many communist regimes).

In the post–Cold War era, environmental activism has become much more diffuse and difficult to pin to larger geopolitical or economic trends. Outspoken activists have been loudly denounced as cranks by conservative regimes in the United States, whereas in Europe and Canada "green" parties have proved quite influential. Japanese critics have not gone unheard, but their voices have been muted lately by economic woes. In much of the developing world, where environmental regulations and enforcement are both lax, activists have faced even greater challenges. In some cases, however, such as that of Choropampa, Peru, site of a 2003 mercury spill related to gold mining (where mercury was being recovered as a by-product of cyanide processing), community outrage over company negligence combined with government corruption led to significant action.

Within this larger context of environmental activism, antimercury activism seems to have gained most traction in recent years by focusing on human health effects caused by everyday exposure, particularly to methylmercury in fish but also to mercury compounds in vaccines and dental amalgam. Activism against mercury pollution from mining, heavy industry, and power generation has remained more specialized and abstract, more difficult for many to link to their or their family's well-being. Although some famous figures have lately become antimercury activists due to their personal travails, activist emphasis on mercury's insidious effects on the health of children and the unborn have been more effective in generating the emotional energy needed to sustain effective political campaigns. In this view, mercury is now unequivocally a public enemy.

## CONCLUSION

Humanity's millennia-long relationship with mercury has been intimate and tangled, torn between attraction and detestation. Since remote antiquity, mercury has yielded many useful pigments, alloys, and medicines. It has enabled recovery of precious metals, as well as the construction of key scientific instruments such as the barometer and thermometer. Mercury compounds have been critical to industry, enabling the manufacture of everything from explosives to batteries to compact fluorescent lightbulbs. In these ways mercury has seemed a remedy and aid to human progress. Yet in keeping with the ambivalent notion of the pharmakon, mercury's toxicity has always been lurking just beyond the light of these innovations, first in the mines and laboratories and factories where countless workers were unwittingly poisoned, but then in the air and water and soil contaminated by vapors and effluents. It is only in the last decades of history, almost the blink of an eye, that humans seem to be coming to grips with the full effects of this dangerous attraction.

## BIBLIOGRAPHY


Acosta, José de. 2002 (1590). *The Natural and Moral History of the Indies*. Jane Mangan, ed. Durham, NC: Duke University Press.

Agricola, Georgius. 1960 (1912). *De re metallica*. Trans. Herbert Hoover. New York: Dover.

Almeida, Paul, and Linda Brewster Stearns. 1998. Political opportunities and local grassroots environmental movements: The case of Minamata. *Social Problems*, 45(1):37–60.

Bakewell, Peter J., ed. 1997. *Mines of Silver and Gold in the Americas*. Aldershot: Variorum.

Bakewell, Peter J. 1971. *Silver Mining and Society in Colonial Mexico, Zacatecas, 1546–1700*. Cambridge: Cambridge University Press, 545–579.

Brading, David, and Harry Cross. 1972. Colonial silver mining: Mexico and Peru. *Hispanic American Historical Review*, 52:545–79.

Brown, Kendall. 2001. Workers' health and colonial mercury mining at Huancavelica, Peru. *The Americas*, 57:467–496.

Cooke, Colin A., Prentis H. Balcom, Harald Biestar, and Alexander P. Wolfe. 2009. Over three millennia of mercury pollution in the Peruvian Andes. *Proceedings of the National Academy of Sciences* 106(22):8830–8834.

Craig, Alan K., and Robert C. West, eds. 1994. *In Quest of Mineral Wealth: Aboriginal and Colonial Mining and Metallurgy in Spanish America (Geoscience and Man 33)*. Baton Rouge: Louisiana State University Press.

Doktor, Raphael. 1938. Pigments of the painter. *Parnassus*, 10(5):21–27.

Eliade, Mircea. 1978 (1956). *The Forge and the Crucible: The Origins and Structures of Alchemy*. 2nd ed. Chicago: University of Chicago Press.

Goldwater, Leonard J. 1972. *Mercury: A History of Quicksilver*. Baltimore: York Press.

Hadsund, Per. 1993. The tin-mercury mirror: Its manufacturing technique and deterioration processes. *Studies in Conservation*, 38(1):3–16.

Harper, Martin. 1987. Possible toxic metal exposure of prehistoric bronze workers. *British Journal of Industrial Medicine*, 44(10):652–656.

Humboldt, Alexander von. 1972 (1811). *Discourse and Reflections on the Kingdoms of New Spain*. Mary Maples Dunn, ed. New York: Alfred A. Knopf.

Lohmann Villena, Guillermo. 1949. *Las minas de huancavelica en los siglos XVI y XVII*. Seville: EEHA.

Michiko, Ishimure. 1990. *Paradise in the Sea of Sorrow: Our Minamata Disease*. Kyoto: Yamaguchi.

Mishima, Akio. 1992. *Bitter Sea: The Human Cost of Minamata Disease*. Tokyo: Kosei.

Pearson, Michael N. 1995. The thin end of the wedge: Medical relativities as a paradigm of early modern Indian-European relations. *Modern Asian Studies*, 29(1):141–170.

Pendergast, David M. 1982. Ancient Maya mercury. *Science (New Series)*, 217(4559):533–535.

Swiderski, Richard. 2008. *Quicksilver: A History of the Use, Lore, and Effects of Mercury*. Jefferson, NC: McFarland & Co.

Whitaker, Arthur P. 1941. *The Huancavelica Mercury Mine: A Contribution to the History of the Bourbon Renaissance in the Spanish Empire*. Cambridge, MA: Harvard University Press.

White, David Gordon. 1984. Why gurus are heavy. *Numen*, 31, Fasc. 1 (July):40–73.

Young, Otis E., Jr. 1970. *Western Mining*. Norman: University of Oklahoma Press.

# 3 Human Impacts on Earth's Natural Mercury Cycle*

*Elizabeth Malcolm*

## CONTENTS



> When we try to pick out anything by itself, we find it hitched to everything else in the Universe.
>
> —**John Muir,** *My First Summer in the Sierra*

## INTRODUCTION

Mercury, a naturally occurring element on the earth, is neither created nor destroyed. Unlike pollutants such as the pesticide DDT made famous by Rachel Carson in *Silent Spring*, or ozone-layer-destroying CFCs, mercury is not manmade. Mercury is found everywhere in small amounts—in rocks, soils, air, water, plants, animals, and humans. It has been continuously moved and cycled throughout the earth over its 4.6-billion-year history both by natural and human processes. However, compared to any natural process, human actions have greatly accelerated movement of this metal out of the earth's crust

* This chapter is dedicated to the memory of Dr. Gerald J. Keeler, my advisor and friend, and international mercury expert. I knew Jerry as an inspiring mentor who could bring out the best in his students. His pioneering contributions in atmospheric mercury research have forever changed our understanding of the mercury cycle and mercury pollution.

where it was stable and sequestered, and transferred it to air, soil, and surface waters where it finds its way into aquatic food webs and, ultimately, into humans.

Even though mercury is a natural element it is still considered a pollutant and has increasingly received a great deal of attention from environmentalists. Why spend billions to study, clean up, and regulate a metal that occurs naturally in the environment? Where does mercury come from, and how does mercury get into our lakes, rivers and oceans? To answer these questions we need to understand the path mercury takes through the environment—how it goes from a volcano or power plant to the air, then water, and into fish; how mercury from a coal plant in China is found in a polar bear in Canada or fish in a New England pond; and how mercury, through a natural processes of bioaccumulation, can prove toxic to wildlife and humans. Mercury is continuously being recycled in this natural, everlasting journey of mercury atoms through living things and the earth (Figure 3.1).

Mercury is global in both its sources and distribution. Natural and industrial emissions of mercury are transported around the world in the atmosphere, resulting in mercury contamination. This chapter examines how humans have affected mercury on a local and global scale and thus increased concentrations of mercury in the environment. What controls mercury's distribution and transport in the environment? To understand the story of mercury's global travels we must understand its chemistry, geology, and biological effects as well as the human impacts on its cycle.

## MERCURY'S CHEMICAL NATURE

Mercury is a heavy metal, "Hg" on the periodic table. It not only shares some properties with its neighboring metals but also has some important and unique characteristics. Its most basic chemical form, elemental mercury, is found by itself and not bonded to any other elements (Benjamin and Honeyman, 1992). This elemental, inorganic mercury is the silver-colored mercury used in thermometers and dental fillings. Unlike other metals it is a liquid at room temperature, making it ideal for many medical and electronic applications.

Like other metals (lead, copper, arsenic, etc.), mercury forms strong bonds with sulfur (S). Hence, most mercury is found in the earth's crust as the mineral cinnabar, HgS, and inside the human body mercury forms bonds with the sulfur groups of proteins (Benjamin and Honeyman, 1992). Mercury can also bond with carbon, forming organic molecules. The most noteworthy of these is called methylmercury. In methylmercury, one or two $CH_3$ ("methyl") groups are bound to mercury. This organic form is the most toxic form of mercury found in the environment, much more toxic than the inorganic forms previously mentioned. Since in its organic form mercury binds to proteins, it markedly accumulates in the tissues of living organisms such as fish. One of mercury's unusual properties is its ability to become a gas at much lower temperatures than all other metals, making it the only metal found as a gas in the earth's atmosphere. As a result, whenever something is heated that contains mercury, such as coal, some of the mercury is released as a gas into the air. These gaseous forms of mercury can then be easily transported in the atmosphere away from an emissions source, traveling even thousands of kilometers. The chemistry of mercury thus makes it a unique pollutant and plays a large role in its use, transport, and toxicity.

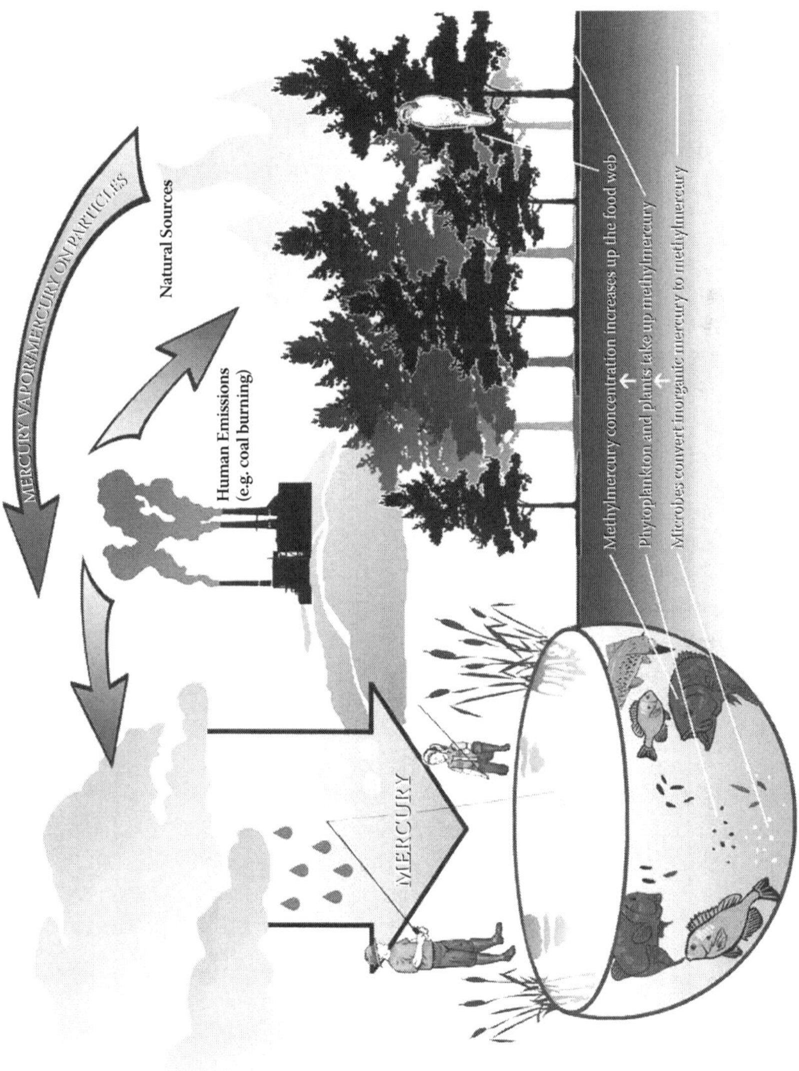


**FIGURE 3.1** The mercury cycle. S. Ryan Lewis ('10), Creative Services, College of William & Mary.

**TABLE 3.1**
**Seafood Species Containing the Highest Concentrations of Mercury as Reported by the US Food and Drug Administration (US FDA, 2009)**

| Species | Mean Concentration (ppm) | Concentration Range (ppm) |
|---|---|---|
| Tilefish | 1.45 | 0.65–3.73 |
| Shark | 0.99 | <0.01–4.54 |
| Swordfish | 0.98 | <0.01–3.22 |
| King Mackerel | 0.73 | 0.23–1.67 |

*Note:* The US FDA recommends that women who may become pregnant, pregnant women, nursing mothers, and young children not consume these fish and limit consumption of other seafood to 12 oz/week.

Due to the toxicity of the chemical form of mercury, methylmercury—even very low concentrations—poses serious health concerns. "One in a million" may not sound like much, but one in a million is the US Food and Drug Administration's (FDA) action limit for mercury in fish.

One milligram of mercury for every million milligrams (parts per million or ppm) of fish is all it takes to be of concern. The FDA found four species of commercial seafood with average mercury concentrations over 1 ppm and recommends that pregnant women not eat these four species: tilefish, swordfish, king mackerel, and shark (see Table 3.1 and Chapter 5).*

The concentrations of mercury that cause health concerns are low, but the concentrations of mercury in the surrounding environment are even lower. However, natural chemical and biological processes concentrate this mercury in organisms. For example, in Minnesota's Lake Tamarack the mercury concentration is only 0.0005 parts per billion (ppb) (Monson and Brezonik, 1998). That amount is equivalent to only 5 grams of mercury for every ten trillion grams of water. Because of a process called biomagnification, the microscopic plankton in Lake Tamarack contains 10 ppb of mercury and the fish have 100 ppb of mercury—one million times more than found in the water. Thus, because of this biomagnification, most people's primary mercury exposure is from fish consumption, not from drinking water. These low concentrations make mercury pollution challenging to study and contribute to the ease of its global distribution. It truly is "a hazard without borders."

## ON THE TRAIL OF MERCURY

As mercury travels through the environment the atoms are constantly changing chemical form and moving among air, water, soil, and biota. Concentrations in the environment are so low that researchers must take care that simple dust and dirt do not contaminate a sample, making the difficult task of tracking mercury pollution

* The advisory is aimed at pregnant women because the developing fetus is ten times more sensitive to mercury, which can damage the developing brain (Clarkson, 1993).

even more challenging. Through decades of research in the field collecting environmental samples and simulating mercury transport and chemistry with computer models, a more complete picture of how mercury travels the globe and is transformed is coming into focus. This global cycling takes mercury through the earth's "spheres": the geosphere (physical earth, soils, rocks), the atmosphere, the hydrosphere (ocean, rivers, lakes, etc.), and the biosphere (living organisms).

## Geosphere

Mercury's most stable reservoir resides in the earth. Mercury is naturally found in some ores, rocks, and soils. As mentioned, the main mineral ore form of mercury is cinnabar, HgS. Mercury can cycle within the geosphere as part of the natural "rock cycle" when sediment is converted to rock or rock to soil. For example, if mercury is present in the underlying bedrock, as the bedrock is weathered and broken apart it enriches the soil with mercury.

Mercury can enter the geosphere through several pathways. Mercury in the atmosphere can enter the soil through wet or dry deposition. Wet deposition is transfer from the atmosphere by precipitation. Dry deposition is the transfer of gases or particles to a surface, such as through gravitational settling of large aerosols. Mercury can also be transferred to soil by the decomposition of organic matter from a biological source. Litterfall is a good example of this—if plant leaves contain mercury (and most do), the mercury is transferred to the soil when the leaves fall and decompose (Mason et al., 2009). Soils in many areas have also been contaminated by human activity either directly, such as through dumping hazardous wastes on the ground, or indirectly through air pollution.

Mercury can then leave the geosphere through natural or anthropogenic (human-caused) processes. In a volcano high temperatures literally melt the rock, and the mercury escapes as a gas. Mercury can escape from soil as a gas, but it does so more slowly. Mercury can also leave the soil as it is washed away through runoff and by rivers or be transferred to the atmosphere as windblown soil particles. Coal contains small amounts of mercury and when extracted from the ground and burned it releases mercury into the atmosphere. Similarly many ores, including gold and copper ores, can contain trace amounts of mercury that are released to the atmosphere when heated during the smelting process to refine the ore (Pirrone et al., 2010).

## Atmosphere

Mercury cycles within the atmosphere in several chemical forms as it is transported with the winds and air circulation. As noted above, mercury can enter the atmosphere in several ways, both natural and human driven. The natural processes that transfer mercury to the atmosphere include volatilization* of mercury from oceans, soil, and vegetation; these processes account for approximately 60% of emissions to the atmosphere, or 5207 mg/year (Figure 3.2) (Pirrone et al., 2010). It is important

* Due to mercury's high vapor pressure, mercury dissolved in water, or present as a solid in soil and on surfaces, can transform to a gas and transfer to the atmosphere.

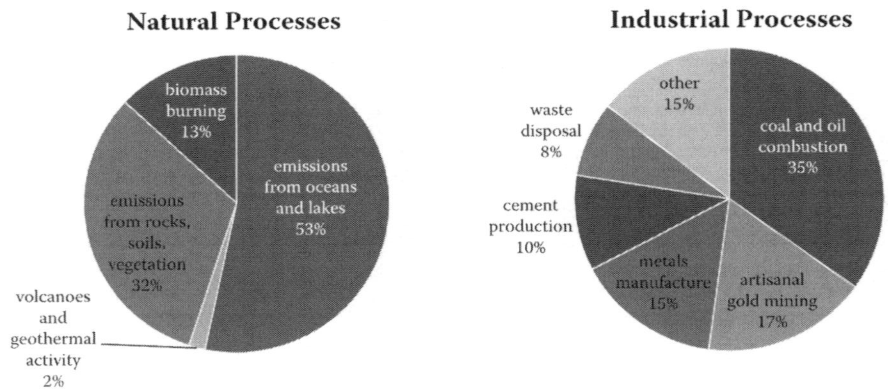


**FIGURE 3.2** Emissions from natural and industrial processes. The "natural" emissions include reemissions of mercury that originally had an anthropogenic origin. Biomass burning includes fires with natural and anthropogenic origin. Data from Pirrone et al., 2010.

to note that this total includes reemission of mercury from human contamination. This reemission occurs because the current volatilization of mercury from surfaces such as the ocean includes both mercury in the ocean with a natural origin and mercury added to the ocean by human activities. Thus, these emissions from natural processes would likely have been lower in preindustrial times, before human influence. The other source of emissions is direct anthropogenic atmospheric emissions, including coal combustion, artisanal gold mining, metals manufacturing, and other industrial processes, all of which account for approximately 2909 mg/year, or 40% of emissions.

The fate of this mercury added to the atmosphere strongly depends on the mercury's chemical and physical form. As a result atmospheric mercury can be transported globally or deposited back to the ground close to the emission source, possibly contributing to local pollution. The most abundant form of atmospheric mercury (>95%) is elemental gaseous mercury ($Hg^0{}_g$), which is not very water soluble and not very reactive, giving it a lifetime in the atmosphere of six months to a year (Lin and Pehkonen, 1999). Thus, mercury emitted in this form can travel very long distances before it is deposited back to the surface. For example, researchers in Oregon have measured mercury in the air that has traveled across the Pacific Ocean (Jaffe et al., 2005). The other forms of atmospheric mercury are much shorter lived, lasting hours to days in the atmosphere. These include mercury bound to particles (soot, sea spray, soil dust, etc.) and what is known as reactive gaseous mercury, a group of gasses that include mercuric chloride, mercuric bromide, and mercury oxide. These forms of mercury are a great concern as local and regional pollution sources. Particulate bound mercury can be deposited to the surface by gravitational settling, incorporation into clouds and precipitation, or impaction with a surface (literally colliding with a surface like a leaf or water as it is carried by the wind). Reactive gaseous mercury is very water soluble so it can easily be incorporated into clouds and precipitation, or it can dissolve into water surfaces that it encounters such as lakes, oceans, and even dew (Malcolm et al., 2002). Gaseous forms of mercury can also be taken up by plants as they take in air for carbon dioxide. Predicting the fate of mercury emissions is further complicated by atmospheric processes, which can convert mercury among these forms.

Bodies of water that have had no direct inputs of mercury have been found to contain fish with elevated mercury levels. In many cases, this mercury has been traced to atmospheric mercury emissions that were transported long distances before deposition to land or water (Expert Panel on Mercury Atmospheric Processes, 1994). Due to atmospheric transport, mercury pollution is truly global, even reaching remote locations such as the Arctic ecosystem (Poissant et al., 2008).

## HYDROSPHERE

This "sphere" includes the world's lakes, rivers, oceans, wetlands, groundwater, and other waterbodies. Mercury is cycled within the hydrosphere as part of the water cycle in both dissolved and particle-bound forms. Most mercury that enters the hydrosphere is in inorganic forms. One of the most important transformations in mercury's journey is methylation, which converts this inorganic mercury to an organic methylated form. As mentioned previously, methylated mercury is much more toxic than inorganic mercury and is the form that accumulates in organisms, including fish. In freshwater and coastal ecosystems, methylation is primarily carried out by bacteria called sulfate reducing bacteria (SRBs). These organisms are anoxic, living in habitats with no oxygen such as lake and wetland sediments. Thus, most methylation does not occur at the oxygenated surface of a lake but when the mercury has made its way down to the sediments. Local chemical conditions, such as the sulfur and organic carbon content of the sediment, affect the overall rate of methylation (Selin, 2009).

The major sources of mercury to surface waters such as lakes and oceans are direct atmospheric deposition and runoff. In many cases, mercury pollution has been added directly to the hydrosphere by human activity. The most famous example is the pollution of Minamata Bay by a local chemical company, Chisso Company (Minamata City, 2000) (see Chapters 1 and 8). The company dumped more than 70 tons of organic mercury waste into the bay beginning in the 1930s. In the 1950s local children began exhibiting severe neurological problems, some died, and incidences of miscarriages and birth defects rose. The cause of this mysterious disease was later determined to be methylmercury waste from the plant that contaminated the local fish and shellfish. Tens of thousands of people have claimed to have been affected by what came to be called "Minamata Disease." This early case of mercury poisoning brought worldwide attention to the dangers of environmental mercury pollution.

Removal processes can transfer mercury from the hydrosphere to the lithosphere, atmosphere, or biosphere. Mercury readily adsorbs to particles in the water and can then settle to the bottom of the lake or waterbody by gravitational settling. This process can transfer mercury to the sediment at the bottom of a lake or ocean where it may quickly cycle back out of sediment or may be essentially trapped in the lithosphere. Some forms of mercury can volatilize out of the water to the atmosphere in the form of a gas via chemical or bacterially mediated reactions. The chemistry of the water and amount of sunlight are important factors in determining the volatilization rate (Mason, 2009). As mentioned above, mercury, particularly methylmercury, can be taken up by organisms and transferred from the hydrosphere into a food web.

## Biosphere

Transfer of mercury within the biosphere is facilitated by its chemical affinity for biological molecules. In aquatic ecosystems, methylmercury not only bioconcentrates from water into the organism but also bioaccumulates, with concentrations increasing in individual organisms over time, and biomagnifies so that concentrations increase as we move up the food web to higher trophic levels. The result is that fish at the top of the food web have the highest concentrations of mercury, much higher than that of the water. Longer-lived individuals will also have more time to accumulate higher concentrations.

New mercury is transferred into the biosphere when organisms take in mercury from their surrounding environment. Terrestrial plants, for example, take up mercury from the soil, atmosphere, and soil water (water residing in the pore spaces between soil particles). Plants also act as important conduits for mercury transfer, moving mercury from the soil water to the atmosphere as they translocate water through the process of transpiration and back from the atmosphere to soil via litterfall (Mason, 2009). As mentioned previously in the example of litterfall, mercury can also leave the biosphere as organisms die and decay. Biomass burning by natural or human causes will also transfer mercury from the biosphere to the atmosphere.

## Lake Champlain Study

An example of how mercury moves through the environment resulted from a long-term study of mercury conducted for the Lake Champlain Vermont Watershed (Figure 3.3). A watershed includes all the land around the lake over which runoff and streams drain into the lake. This watershed is primarily rural, without many direct sources of mercury pollution. Yet the fish and sediments of the lake have high concentrations of mercury (McIntosh, 1994). In this long-term study, researchers collected and analyzed samples from all components of the environment—air, water, leaves, even clouds—to investigate how atmospheric mercury enters the lake (see, for example, Scherbatskoy et al., 1998; Malcolm et al., 2003; Keeler et al., 2005; and Gao et al., 2006). What they found is that mercury primarily enters the watershed and lake by streams' input (56%), followed by direct atmospheric deposition to the lake surface (20% wet deposition, 18% dry deposition), and wastewater treatment plant discharges (5.7%) (Gao et al., 2006). Since Lake Champlain has a large watershed

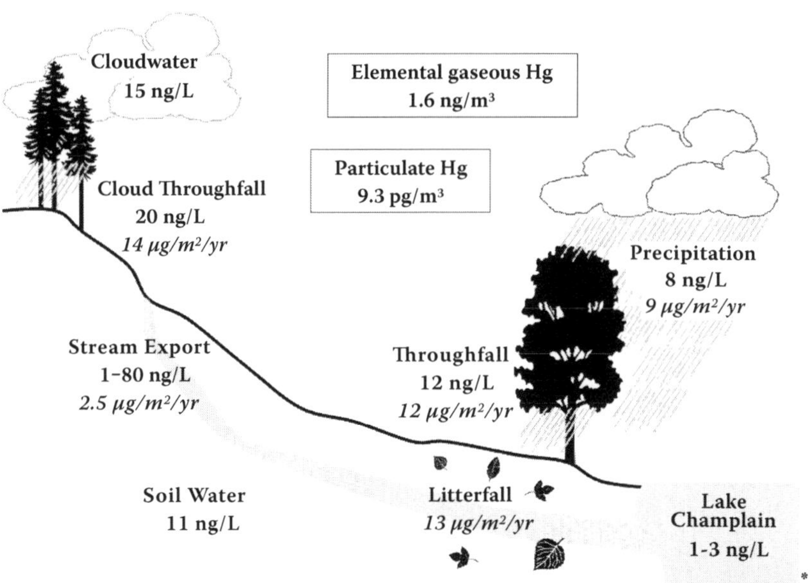


**FIGURE 3.3** Results of the Lake Champlain mercury study illustrating the pathways of mercury from the atmosphere to the watershed. Aqueous concentrations are in ng/L. Annual amount of mercury transferred by each process (in italics) is in μg/m²/yr. Data from Scherbatskoy et al., 1998; Lawson et al., 2003; and Malcolm et al., 2003.

area compared to the size of the lake (18:1) and few point sources of mercury, most of the mercury in streams is also derived from atmospheric deposition. The wet deposition is predominantly from rain and snow, and the dry deposition includes direct transfer of mercury gases and aerosols to surfaces such as the lake and tree canopy. Cloud water deposition on trees on Mt. Mansfield also contributed mercury to the watershed since clouds had nearly twice the mercury concentration of rain (Malcolm et al., 2003). Precipitation or cloud drip measured under a tree canopy is termed "throughfall" and is typically much higher in mercury than the original precipitation or cloud water (Figure 3.3). This difference is primarily due to incorporation of dry deposited mercury as the precipitation washes off the leaves, and demonstrates the importance of dry deposition to a forested watershed. As seen in Figure 3.3, much of the mercury deposited to a forested watershed, like that of Lake Champlain, is retained by the soil, and some is reemitted to the atmosphere. The mercury bound to soil can slowly wash into the lake via streams and runoff. Thus, there will be a lag time in the response of ecosystems to changes in atmospheric deposition. Knightes et al. (2009) found that fish concentrations responded more slowly in lakes with large watersheds versus lakes with small watersheds, which would receive most mercury input from direct atmospheric deposition to the lake.

## HUMAN PERTURBATION

Today we know that humans have affected nearly all aspects of the environment; from species extinction to global warming, human influence reaches all corners of the globe. Mercury is no exception. The oceans and atmosphere seem so vast that people once thought "the solution to pollution was dilution." So if the smoke stack was tall enough, the soot and gases would dissipate in the atmosphere. However, smoke stacks now clutter urban skylines, concentrating pollutants, and we now know that many pollutants persist in the environment, lasting months, years, or even decades.

Mercury is truly a global pollutant with emissions coming from around the globe (Figure 3.4). Europe and North America were once the largest emitters of mercury. Environmental regulations have reduced emissions from Europe and North America over the last few decades while emissions from industrializing countries have been growing; yet the total amount of global emissions has remained nearly constant (Pirrone et al., 2009; Pirrone et al., 2010). Recent estimates show large increases in emissions, corresponding to rapid economic development from Asia, particularly China and India, as well as Africa.

Scientists have tried to go back in time to see what the environment looked like before human perturbation and to estimate the anthropogenic effects on mercury concentrations. In the 1990s Drs. Mason, Fitzgerald, and Morel asked just these questions on a global scale (Mason et al., 1994). How much mercury was in each component, or "reservoir" of the earth's system before human influence, and how much is there today? Some of these numbers are known with relative certainty, some are highly uncertain, but the exercise demonstrates a valuable point—human activity has greatly changed the amount of mercury cycling through the air and water and therefore the amount to which organisms are exposed. Results of their analysis are summarized below (Figure 3.5). The other major point this study demonstrated is that changes in

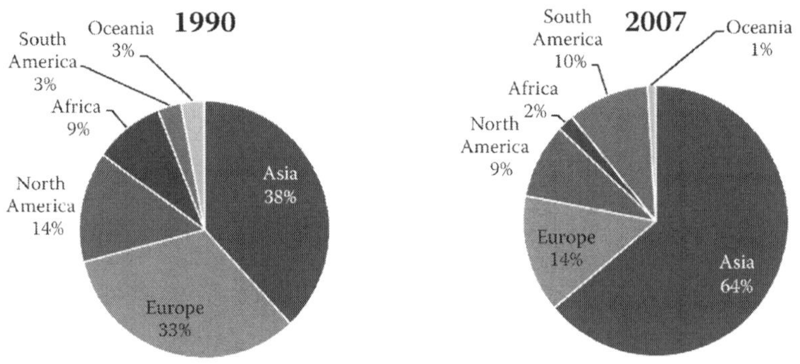


**FIGURE 3.4** Estimates of global anthropogenic (human-caused) emissions of mercury to the atmosphere by region for the years 1990 and 2007. Although emission sources have changed, the total global emissions were estimated to differ by less than 5% between 1990 and 2007. Redrawn from Pirrone et al., 2010.

one component of the earth's system have a cascading effect on other parts of the cycle. For example, humans have increased emissions of mercury to the atmosphere, which in turn increased the amount of mercury that comes back to the surface in precipitation and dry deposition. Thus, our actions are not isolated. The earth is a system with interrelated components that are all connected. Mercury emissions in industrialized areas and cities persist and are then transported in the atmosphere and rivers to "pristine" ecosystems. On a global scale, humans have increased the amount of mercury cycling in every component of the earth's system.

Just how much have humans impacted the mercury cycle? Current estimates suggest that the global average concentration of mercury in the lower layer of the atmosphere (troposphere) has increased by three to five times since preindustrial levels (Selin, 2009). Data on ocean impacts are scarce, so the exact effect of humans has

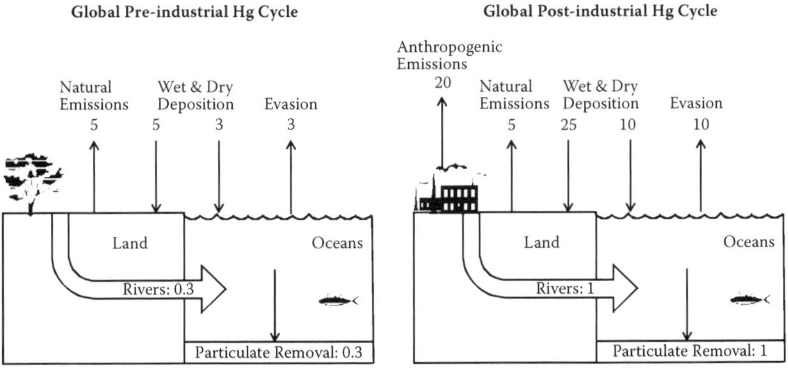


**FIGURE 3.5** On the left is the preindustrial Hg cycle. On the right is the postindustrial Hg cycle. Note that all the fluxes are estimated to have increased due to human activities. Redrawn from Mason, Fitzgerald, and Morel, 1994.

not been comprehensively quantified. Humans primarily affect the ocean surface waters via atmospheric deposition, riverine inputs, and runoff. The ocean is vast and circulates much more slowly than the atmosphere, thus it takes decades for inputs at the surface to mix through the whole ocean. Although mercury concentrations throughout the environment have changed globally, the change is not uniform. Areas closest to emission sources have the highest mercury concentrations in the environment and often have fish with dangerously high concentrations. Thus, the impacts of humans on the mercury cycle are global in their reach, but can be most pronounced on local scales.

## THE OCEAN FRONTIER

The story of mercury's path through the atmosphere to freshwater fish is well understood. But what is the source of mercury in the commercial seafood we all eat? How much is natural and how much is added by human activity? Recent studies shed light on this question, but debate is ongoing and much is unknown. Pollution initially reaches coastal and surface waters, so mercury concentrations in these regions are the first impacted by humans. Due to the great depth of the ocean and its slow circulation, pollution from atmospheric deposition and river runoff takes time to spread throughout the ocean. In coastal and freshwater systems, sulfate reducing bacteria (SRB) methylate mercury in anoxic sediments. However, in the open ocean food webs fish and other organisms often live high above the anoxic sediments where SRB could produce methylmercury. To further add to the mystery, dimethyl mercury, which contains two methyl groups, is commonly found in marine waters but not freshwater (Fitzgerald et al., 2007). This dimethylmercury may be formed by a mechanism unique to marine waters and then degraded to the longer-lived monomethymercury, which then accumulates in fish (Mason and Fitzgerald, 1993). The source of these forms of methylated mercury in the open ocean, far from coasts, is unknown. This knowledge is crucial for determining how human mercury pollution affects mercury concentrations in seafood such as tuna and swordfish. Several hypotheses have been proposed and are being investigated including bacterial methylation in deep ocean sediments, bacterial or chemical methylation in hydrothermal vents, bacterial methylation in surface or low-oxygen waters, and transport of methylmercury by currents or the food chain from coastal areas. In the first two hypotheses the source of the methylated mercury is natural. Deep-ocean sediment is not yet expected to be greatly contaminated by human mercury sources, and many hydrothermal vents have been found to naturally release mercury from the deep earth into the ocean. For the latter two hypotheses, however, humans have increased total mercury available for methylation.

Since 2000 researchers have begun to investigate the question of where open ocean methylmercury originates through field measurement, experiments, and model simulations. Recent studies provide direct or indirect evidence that mercury can be methylated in oxygenated marine waters, possibly by bacteria that decompose dead plankton and small fecal pellets as they sink through the water column (Monperrus et al., 2007; Kirk et al., 2008; Cossa et al., 2009; Conway et al., 2009; Sunderland et al., 2009). Continued research on this topic will allow scientists to determine where mercury in popular marine seafood species originates and inform policymakers on what impact anthropogenic emissions reductions would have on the mercury content of seafood.

## MERCURY FOOTPRINT

Humans seldom consider the impact of their daily lives on the problem of global mercury pollution. Although we as individuals may not directly release mercury into the air or water, we all contribute indirectly to emissions. Any activity dependent on electricity generated from coal combustion will add to atmospheric emissions and the transfer of mercury from the geosphere to the atmosphere. This activity may be direct electricity usage in our homes and work but could also include the products we purchase that were made using electricity generated from coal. The chemical properties of mercury have led to its widespread use, particularly in electronics and medical applications (Table 3.2). Upon disposal, products containing mercury that are incinerated will add to air emissions, and improperly disposed products could add to water and air pollution. In the United States, most localities and states have programs for hazardous waste collection that recycle mercury from products, like thermometers and compact fluorescent bulbs, and properly dispose of other mercury products like paint in hazardous waste landfills (US Environmental Protection Agency, 2010a). Modern hazardous and municipal waste landfills are now designed to prevent any transfer of mercury and other waste components from leaching into groundwater and surface waters (US Environmental Protection Agency, 2008).

**TABLE 3.2**
**A Partial List of Common Mercury-Containing Products (US Environmental Protection Agency, 2010b)**

| Product | Purpose of Mercury |
|---|---|
| Liquid crystal display (LCD) monitors and screens | Emits light when energized |
| Fluorescent bulbs, ultraviolet lights, neon lights | Emits light when energized |
| Old washing machines, dryers, chest freezers, clothes irons, televisions, silent light switches, automobiles | May contain tilt switches, in which liquid mercury in a tube flows to make an electrical connection when the tube is tilted; most uses have been discontinued |
| Thermostats | Temperature sensitive mercury tilt switches |
| Alkaline, zinc air, and silver oxide button cell batteries | Prevents degradation and rupture of battery |
| Thermometers, barometers | Silver-colored liquid mercury expands and contracts in response to temperature and pressure changes |
| "Silver" dental amalgam fillings | Binds metals together into amalgam |
| Blood pressure cuffs | Liquid mercury in tube responds to blood pressure changes |

Ecological footprints and carbon footprints were developed as a way to estimate an individual's impact on the environment overall or through greenhouse gas emissions, respectively. A carbon footprint, for example, accounts for an individual's greenhouse gas emissions from transportation, electricity, food, and consumer purchases. By applying the same concept to mercury, an individual or organization's "mercury footprint" can be calculated. The largest contribution to most individuals' mercury footprints comes from electricity generation, more specifically electricity generated from coal (see Figure 3.2). Other fossil fuels, such as oil and natural gas emit negligible amounts of mercury by comparison (Pirrone et al., 2009). The emissions of mercury from coal combustion will vary by power plant dependent on the mercury content of the coal burned, the efficiency of electricity generation at the plant, and environmental controls designed to reduce mercury emissions. The Leonardo Academy (2010) has created a mercury footprint calculator based on annual electricity use. On their interactive website (http://www.cleanerandgreener.org/resources/pollutioncalculator.html), a user can choose their state of residence and enter their annual kilowatt-hours of electricity usage to estimate their annual mercury emissions. A simple worksheet to calculate mercury emissions from electricity for an average American using this website's method is presented in the box below.

**WHAT IS MY MERCURY FOOTPRINT?**

The following calculation estimates your mercury emissions to the atmosphere by considering your annual electricity usage and the average US electricity emissions profile. For emissions tailored to your state see http://www.cleanerandgreener.org/resources/pollutioncalculator.html.

_______ kwh of annual electricity × 0.013066 = ______ mg mercury emitted

Environmental footprints are also estimated by calculating the environmental impact on a per capita basis. This method not only takes into account an individual's direct impact such as through electricity and water usage at home, but also accounts for impacts from their larger community. Although such an estimation of a mercury footprint does not account for individual differences in mercury emissions from person to person, it does account for those indirect emissions we are all responsible for, such as electricity used by schools, government buildings, and businesses, and mercury emitted from industrial production and waste incineration. An estimate of the mercury emissions per capita is presented in Table 3.3 and was calculated by dividing the total Hg emissions by the population for the year 2005. Although these numbers have a large associated uncertainty due to uncertainty in emissions* and population data, regional comparisons are interesting. For example, although Asia accounted for more than 50% of the estimated mercury

* Emissions estimates may have 30%–50% associated uncertainty per emission type (Pirrone et al., 2010).

**TABLE 3.3**
**An Estimate of Mercury Emission Per Capita**

| | Total Hg Emissions[a] Mg/Yr | Population[b] In Thousands | Emissions Per Capita g/Yr |
|---|---|---|---|
| South Africa[c] | 40.2 | 48,073 | 0.84 |
| China[d] | 609.1 | 1,312,253 | 0.46 |
| India[c] | 240.9 | 1,130,618 | 0.21 |
| Australia[e] | 16.6 | 20,395 | 0.81 |
| Europe[e] | 145.2 | 729,421 | 0.20 |
| Russia[e] | 69.8 | 143,170 | 0.49 |
| North America[e] | 152.8 | 335,175 | 0.46 |
| South America[e] | 49.7 | 371,658 | 0.13 |
| World Total | 2319.7 | 6,512,276 | 0.36 |

[a] Data from Pirrone et al., 2010.
[b] Data from United Nations Population Division, 2010; reference year for data used was 2005.
[c] Reference year for emissions data was 2004.
[d] Reference year for emissions data was 2003.
[e] Reference year for emissions data was 2005.

emissions in 2007, the per capita emissions for China and India were comparable to other regions.

If some or all of your electricity comes from coal, reductions in energy usage would reduce your overall mercury footprint. However, one of the most common energy efficiency products, compact fluorescent light bulbs, contains mercury. So would using compact fluorescent lamps (CFLs) increase or decrease your mercury footprint? Eckelman and colleagues addressed this question by assessing the mercury emissions from CFLs from manufacture through disposal versus their reduction in mercury emissions via reduced energy demand (Eckelman et al., 2008). Mercury is primarily emitted from CFLs when they are broken during transport or disposed of via landfills and incinerators. When bulbs are recycled properly, however, little mercury is lost to the atmosphere and most is recovered. For the United States about 20% of CFLs are recycled. For states and countries that rely heavily on coal for power, Eckelman et al. found that increased use of CFLs would reduce total mercury emissions. Reduced mercury emissions due to CFLs was predicted to be highest for North Dakota, West Virginia, New Mexico, and the countries China, Kazakhstan, Botswana, Romania, Bulgaria, Republic of Macedonia, Greece, and Estonia. Interestingly, increased use of CFLs were predicted to increase mercury emissions in Alaska, California, Oregon, Idaho, Vermont, New Hampshire, Maine, and Rhode Island—all states with little reliance on coal for energy. Similarly, in countries with low coal usage and little to no CFL recycling, increasing CFLs would increase mercury emissions. This scenario was predicted for several countries in Latin America, Africa, and the Middle East. For the United States as a whole shifting to CFLs would produce a net mercury emission reduction. It is worth noting that the reductions in energy from CFL use can have additional benefits by reducing emissions of other pollutants from fossil fuel combustion such as greenhouse gases, nitrogen oxides, and sulfur dioxide.

In addition to reducing your direct use of coal-derived energy, you can lower your mercury footprint by reducing your indirect usage of coal energy—that energy used to make products purchased in China and other countries with high rates of mercury emissions from coal. Buying fewer goods overall and a higher proportion of secondhand items also has the benefit of reducing your overall ecological footprint by reducing your emissions of greenhouse gases, water use, and waste produced.

Lastly, recycling or properly disposing of mercury-containing products will reduce their release to air and water. Products from which mercury can be recycled include the aforementioned CFLs, mercury thermometers, mercury-containing thermostats, and mercury tilt switches. In the United States many mercury-containing products have been phased out including mercury in paint and tilt switches for household appliances. The mercury content has been reduced in many other products for which no suitable mercury replacement exists such as CFLs and button cell batteries (US Environmental Protection Agency, 2010b).

## CONCLUSION

The global distribution of mercury sources and the environmental fate of mercury are dependent on both biogeochemistry and human action. Natural and human processes transfer mercury from the stable reservoir of the earth's crust to the atmosphere and hydrosphere, where mercury can easily travel around the world. Once in an aquatic ecosystem, mercury can be methylated to its most toxic form, which then accumulates in fish and other organisms. Around the world we can find seafood species with high concentrations of mercury, even in pristine ecosystems. Humans have increased the cycling of mercury through the environment via coal-fired power plants, artisanal gold mining, and other anthropogenic processes. Although these emissions are decreasing in some developed countries due to government regulation, emissions from developing countries are increasing. Even if anthropogenic emissions are substantially reduced globally, legacy pollution will continue to cycle through the environment for centuries. Thus, mercury not only crosses political boundaries but generational ones as well to impact ecosystems and peoples distant in space and time from its source.

## REFERENCES

Benjamin, Mark M., and Bruce D. Honeyman. 1992. Trace metals. In *Global Biogeochemical Cycles*. Samuel S. Butcher, Robert J. Charlson, Gordon H. Orians, and Gordon V. Wolfe., eds. 317–52, London: Academic Press.

Clarkson, Thomas W. 1993. Mercury—Major issues in environmental-health. *Environmental Health Perspectives,* 100:31–38.

Conaway, Christopher H., Frank J. Black, Melanie Gault-Ringold, J. Timothy Pennington, Francisco P. Chavez, and A. Russell Flegal. 2009. Dimethylmercury in coastal upwelling waters, Monterey Bay, California. *Environmental Science and Technology,* 43:1305–9.

Cossa, Daniel, Bernard Averty, and Nicola Pirrone. 2009. The origin of methylmercury in open Mediterranean waters. *Limnology and Oceanography,* 54:837–44.

Eckelman, Mathew J., Paul T. Anastas, and Julie B. Zimmerman. 2008. Spatial assessment of net mercury emissions from the use of fluorescent bulbs. *Environmental Science and Technology,* 42:8564–8570.

Expert Panel on Mercury Atmospheric Processes. 1994. *Mercury Atmospheric Processes: A Synthesis Report. Electric Power Research Institute Report No.TR-104214.* EPRI, Palo Alto, CA.

Fitzgerald, William F., Carl H. Lamborg, and Chad R. Hammerschmidt. 2007. Marine biogeochemical cycling of mercury. *Chemical Reviews,* 107:641–662.

Jaffe, Daniel, Eric Prestbo, Phillip Swartzendruber, Peter Weiss-Penzias, Kato Shungo, Akinori Takami, Shiro Hatakeyama, and Kajii Yoshizumi. 2005. Export of atmospheric mercury from Asia. *Atmospheric Environment,* 39:3029–3038.

Keeler, Gerald J., Lynne E. Gratz, and Khalid Ali-Wali. 2005. Long-term atmospheric mercury wet deposition at Underhill, Vermont. *Ecotoxicology,* 14:71–83.

Kirk, Jane L., Vincent L. St. Louis, Holger Hintelmann, Igor Lehnherr, Else Brent, and Laurier Poissant. 2008. Methylated mercury species in marine waters of the Canadian high and sub Arctic. *Environmental Science and Technology,* 42:8367–8373.

Knightes, Christopher D., Elsie M. Sunderland, M. Craig Barber, John M. Johnston, and Robert B. Ambrose Jr. 2009. Application of ecosystem-scale fate and bioaccumulation models to predict fish mercury response times to changes in atmospheric deposition. *Environmental Toxicology and Chemistry,* 28:881–893.

Lawson, Sean T., Timothy D. Scherbatskoy, Elizabeth G. Malcolm, and Gerald J. Keeler. 2003. Cloud water and throughfall deposition of mercury and trace elements in a high elevation spruce fir forest at Mt. Mansfield, Vermont. *Journal of Environmental Monitoring,* 5:578–583.

Leonardo Academy, Emissions Calculator. 2010. http://www.cleanerandgreener.org/resources/pollutioncalculator.html.

Lin, Che Jen and Simo O. Pehkonen. 1999. The chemistry of atmospheric mercury: A review. *Atmospheric Environment,* 33:2067–2079.

Malcolm, Elizabeth G., and Gerald J. Keeler. 2002. Measurements of mercury in dew: Atmospheric removal of mercury species to a water surface. *Environmental Science and Technology,* 36:2815–2821.

Malcolm, Elizabeth G., Gerald J. Keeler, Sean T. Lawson, and Timothy D. Scherbatskoy. 2003. Mercury and trace elements in cloud water and precipitation collected on Mt. Mansfield, Vermont. *Journal of Environmental Monitoring,* 5:584–590.

Mason, Robert P. 2009. Mercury emissions from natural processes and their importance in the global mercury cycle. In *Mercury Fate and Transport in the Global Atmosphere: Emissions, Measurements and Models,* Nicola Pirrone and Robert Mason, eds., 173–191. New York: Springer.

Mason, Robert P., and William F. Fitzgerald. 1993. The distribution and biogeochemical cycling of mercury in the Equatorial Pacific Ocean. *Deep Sea Research I,* 40:1897–1924.

Mason, Robert P., William F. Fitzgerald, and Francois M.M. Morel. 1994.The biogeochemical cycling of elemental mercury—anthropogenic influences. *Geochimica et Cosmochimica Acta,* 58:3191–3198.

McIntosh, Alan, ed. 1994. *Lake Champlain Sediment Toxics Assessment Program: An Assessment of Sediment Associated Contaminants in Lake Champlain, Phase I, Lake Champlain Basin Program Technical Report No. 5.* Grand Isle, VT: Lake Champlain Basin Program.

*Minamata Disease: Its History & Lessons.* 2000. Minamata, Japan: Minamata City.

Monperrus, M., E. Tessier, D. Amouroux, A. Leynaert, P. Huonnic, and O.F.X. Donard. 2007. Mercury methylation, demethylation and reduction rates in coastal and marine surface waters of the Mediterranean Sea. *Marine Chemistry,* 107:49–63.

Monson, Bruce A., and Patrick L. Brezonik. 1998. Seasonal patterns of mercury species in water and plankton from softwater lakes in Northeastern Minnesota. *Biogeochemistry,* 40:147–162.

O'Driscoll, Nelson J., Andrew Rencz, and David R.S. Lean. 2005. The biogeochemistry and fate of mercury in the environment. In *Metal Ions in Biological Systems, Volume 43: Biogeochemical Cycles of Elements*, Astrid Sigel, Helmut Sigel, and Roland K.O. Sigel, eds., 221–238. New York: Taylor and Francis.

Pirrone, Nicola, Sergio Cinnirella, Xinbin Feng, Robert B. Finkelman, Hans Friedli, Joy Leaner, Rob Mason, Arun B. Mukherjee, Glenn Stracher, David G. Streets, and Kevin Telmer. 2009. Global mercury emissions to the atmosphere from natural and anthropogenic sources. In *Mercury Fate and Transport in the Global Atmosphere: Emissions, Measurements and Models*, Nicola Pirrone and Robert Mason, eds., 1–47. New York: Springer.

Pirrone, Nicola, Sergio Cinnirella, Xinbin Feng, Robert B. Finkelman, Hans R. Friedli, Joy Leaner, Rob Mason, Arun B. Mukherjee, Glenn B. Stracher, David G. Streets, and Kevin Telmer. 2010. Global mercury emissions to the atmosphere from anthropogenic and natural sources. *Atmospheric Chemistry and Physics,* 10:5951–5964.

Poissant, Laurier, Hong H. Zhang, João Canario, and Phillipe Constant. 2008. Critical review of mercury fates and contamination in the Arctic tundra ecosystem. *Science of the Total Environment,* 400:173–211.

Scherbatskoy, Timothy, James B. Shanley, and Gerald J. Keeler. 1998. Factors controlling mercury transport in an upland forested catchment. *Water, Air, and Soil Pollution,* 105:427–438.

Schroeder, William H., and John Munthe. 1998. Atmospheric mercury—An overview. *Atmospheric Environment,* 32:809–822.

Selin, Noelle E. 2009. Global biogeochemical cycling of mercury: A review. *Annual Review of Environment and Resources,* 34:43–63.

Sunderland, Elsie M., David P. Krabbenhoft, John W. Moreau, Sarah A. Strode, and William M. Landing. 2009. Mercury sources, distribution, and bioavailability in the North Pacific Ocean: Insights from data and models. *Global Biogeochemical Cycles,* 23:GB2010, doi:10.1029/2008GB003425.

United Nations Population Division. 2010. World Population Prospects: The 2008 Revision Population Database. http://esa.un.org/unpp/.

US Environmental Protection Agency. 2008. Wastes. http://www.epa.gov/wastes/index.htm.

US Environmental Protection Agency. 2010a. State and Local Mercury Collection/Recycling/ Exchange Programs. http://www.epa.gov/wastes/hazard/tsd/mercury/collect.htm.

US Environmental Protection Agency. 2010b. Table of Products that May Contain Mercury and Recommended Management Options. http://www.epa.gov/ wastes /hazard /tsd/ mercury/con-prod.htm.

US Food and Drug Administration. 2004. What You Need to Know About Mercury in Fish and Shellfish. http://www.fda.gov/Food/FoodSafety/Product-SpecificInformation/Seafood/ FoodbornePathogensContaminants/Methylmercury/ucm115662.htm.

US Food and Drug Administration. 2009. Mercury Levels in Commercial Fish and Shellfish. http://www.fda.gov/Food/FoodSafety/Product-SpecificInformation/Seafood/ FoodbornePathogensContaminants/Methylmercury/ucm115644.htm.

# 4 Mercury by the Numbers

*Michael C. Newman and Kenneth M. Y. Leung*

## CONTENTS



It is therefore worth while to search out the bounds between opinion and knowledge.

—**John Locke,** *An Essay Concerning Human Understanding*

## INTRODUCTION

### HARMFUL AMOUNTS OF MERCURY? BOUNDING OPINION AND KNOWLEDGE

Most personal decisions about how best to act are based on opinions of others, not rigorous examination of facts by oneself. Our world is too complex for each of us to examine exhaustively, assess, and then decide about every important issue. Such a tactic would soon result in dysfunctional indecision. Since all opinions do not emerge from the same amount of thoughtfulness and objectivity, the key to making wise decisions is understanding the nature of each opinion and then gauging how well it reflects practical knowledge. Making decisions about environmental issues is particularly difficult, but nonetheless very important to our well-being. Environmental issues permeate our laws, politics, economics, personal health, and collective ethos.

Mercury contamination is arguably one of the most prominent environmental concerns facing us today. Our collective attention was drawn to mercury in the late 1950s when an outbreak of human poisonings occurred in Minamata, Japan (Smith and Smith, 1975). Mercury has kept our attention since then, accumulating an overburden of fact and opinion. These facts and opinions are applied to many decisions. Should one avoid having his or her child vaccinated for the H1N1 influenza because the mercury-based thimerosal preservative in some vaccines has been linked to autism? Should one reduce mercury in his or her diet by avoiding seafood that is also an excellent source of healthy omega-3 fatty acids? Does the mercury in the new low-energy lightbulbs obligate boycott of this product? Winnowing fact from opinion is crucial because mercury can present a real risk* to people and valued wildlife under specific circumstances.

Gathering sound facts and using them effectively to gauge risk is difficult without formal training. Without training the layperson resorts to what cognitive psychologists call informational mimicry (Richerson and Boyd, 2005; Vernimmen et al., 2005), that is, taking the position of an expert whom they trust for some reason such as past reliability or homophily.† This approach is sensible because mimicry is the best strategy when learning facts oneself is error prone and decision error carries substantial cost (Richerson and Boyd, 2005). Unfortunately the most accessible sources of environmental information are now Internet sites that differ widely in reliability and television news programs written as much to entertain as to inform. When confronted with such diverse and confounded information the opinion of most people tends to rely heavily on homophily. Information is sought during decision-making from comfortable and "credible" opinion leaders, and then explored with friends and neighbors with the intent of getting validation for one's emerging decision (Kasperon and Kasperson, 1996). To minimize postdecisional stress, the new opinion is then defended ad hoc from any new information. Unfortunately this common strategy is unreliable because the information for making important decisions about mercury varies so widely in quality and the reliance on homophily is excessive.

## THE ADVANTAGE OF MEASURING

> A precise statement can be more easily refuted than a vague one, and it can be better tested. This consideration also allows us to explain the demand that qualitative statements should if possible be replaced by quantitative ones.
>
> **—Popper (1972)**

Given the above challenge, what is the best way to gather information about environmental mercury? The more explicitly a fact can be stated, the more easily its mettle is tested and the more unambiguous predictions can be made from it. Since quantitative facts or statements are easier to assess and use than are qualitative ones, the intent here is to delve into quantitative measurement and prediction of mercury's potential effects.

### SOME CASES IN WHICH MEASUREMENT WAS UNNECESSARY

An argument could be made that demanding careful attention to measurement is unnecessary in many cases such as judging unacceptable the risk imposed by mercury-

* Risk is the probability or chance of a specific harmful effect or event occurring, for example, a one out of 100 chance of lung cancer if one is a heavy smoker.

† Homophily is the degree to which two or more interacting individuals are similar in relevant features such as education, professional affiliation, political beliefs, or religious upbringing.

tainted fish eaten by Minamata citizens in the 1950s through 1970s (Smith and Smith, 1975). Irreparable neurological damage occurred to young and old, leading to the current resounding conclusion of unacceptable risk. Yet even in this instance twelve years passed between the first recorded case of *in utero* mercury poisoning (1956) and when mercury was officially identified as the cause of the Minamata Disease outbreak (1968) (Smith and Smith, 1975). Minamata residents continued to debate the cause well after publication of the famous Minamata photographic exposé by Smith and Smith (1975). Therefore, even in the retrospectively most obvious of instances, quantification could have accelerated acceptance of the fact and motivated quicker risk reduction.

## Everyday Cases in Which Measurement Is Necessary

Many instances exist in which measurement is desirable or needed to decide wisely about pollutant risk. Measurement can even provide an overarching context for discussing pollutant-related dangers. As one example from two decades ago, Figure 4.1 suggests that voluntary risks to life from tobacco use, poor diet, sedentary lifestyle, and alcohol abuse were much higher in the United States than that from all toxic agents

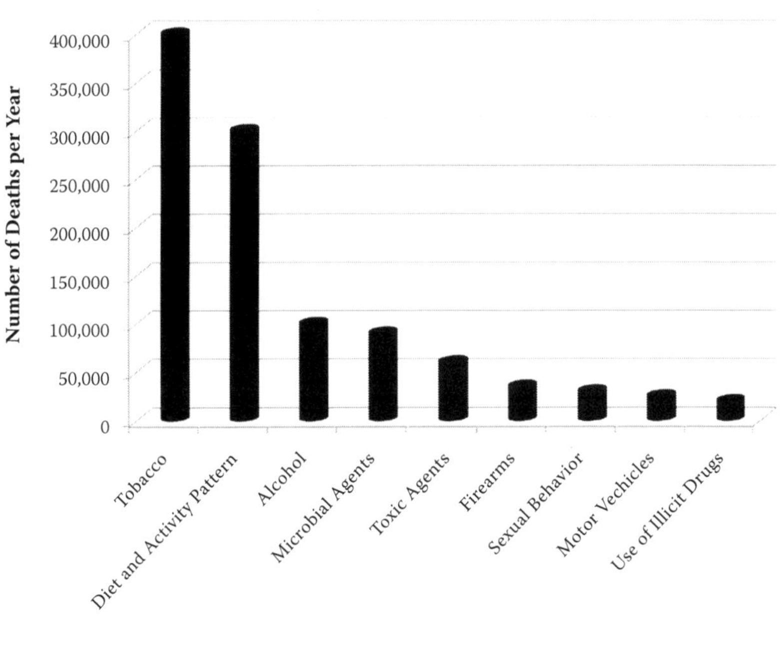


**FIGURE 4.1** The most prominent root causes of death in the United States (1990). Deaths are expressed as simple numbers of deaths that year. These contributors account for approximately one-half of all deaths that year, i.e., Tobacco (19%), Diet and Activity Pattern (14%), Alcohol (5%), Microbial Agents (4%), Toxic Agents (3%), Firearms (2%), Sexual Behavior (1%), Motor Vehicles (1%), and Use of Illicit Drugs (<1%). Toxic-agent-related deaths include those associated with occupational exposures and consumer products in addition to deaths from environmental pollutants. Data from McGinnis and Foege (1993).

combined. Yet toxic agent exposure was and still is perceived as much less acceptable by Americans because it is an involuntary risk: One does not choose to take on toxic risk in order to reap the benefits of the technology that produced the toxicants.

Generally risks taken voluntarily by individuals are perceived as more acceptable than involuntary risks of equal seriousness (Gigerenzer, 2002). Acceptability of risk is not determined by objective fact alone: Many social and psychological factors influence a perceived risk's seriousness or acceptability. These include the ease with which the danger is visualized (Piatelli-Palmarini, 1994), past trustworthiness of any related institution (Beamish, 2001), an individual's openness to new technologies (Petrie et al., 2005), an individual's awareness of environmental issues (Gigerenzer, 2002; Winters et al., 2003), and the manner in which a risk situation is presented initially (Gigerenzer, 2002). Clearly the exposure to the toxicant and also any effect resulting from exposure must be measured or predicted quantitatively to minimize muddling misperceptions.

## FRAMING THE QUESTION

It is also critical to understand the vantage from which a risk is being judged. Perhaps the best means of illustrating this point is to consider the belief that seat belts save lives. This pervasive belief is sound from the vantage of a person in a serious accident whose risk of death does decrease with seat belt use. But seat belts do not save lives from the vantage of the total number of vehicular fatalities in a country. According to British traffic expert John Adams, people wearing seat belts tend to feel safer and consequently drive less cautiously, resulting in more deaths to individuals in other vehicles or pedestrians struck by the less cautious, belted drivers (Adams, 1995). Thus, seat belts do save the lives from the vantage of those wearing them but do not save lives relative to the entire population of a country. As another example, the importance of vantage was obvious recently when proposed changes in screening for breast, prostate, and colon cancers generated an outburst of confusion in the United States, e.g., Marshall (2010). Epidemiologists had concluded that less intense screening protocols were the most sensible and cost-effective ways to address these cancers in the population. This announcement was met with suspicion based on testimonials from people who, without early detection through screening, would have had a much higher risk of dying. These contrasting decisions were a consequence of clashing vantages.

From the vantage of the health care community, some of the money spent in the current cancer screening regimes would be better spent on other important health risks. Using prostate cancer as an example, Anriole et al. (2009) found no difference in death rates from prostate cancer between groups differing in the diligence of prostate-specific antigen and digital rectal examination screening regimes. So from the vantage of managing the risk of death for the entire population from all diseases, the conventional screening is best replaced by the less intense screening.*

* Complicating the issue further, Schroder et al. (2009) found a 20% lowered death rate with prostate specific antigen (PSA)-based screening for prostate cancer but noted that this improvement came at the elevated risk of overdiagnosis. Overdiagnosis carries consequences such as septicemia or excessive bleeding after biopsy.

Equally valid were the conclusions in testimonies of individuals for whom early detection greatly decreased their personal risk of death from cancer. Risk of death was reduced substantially for the individual who actually had the cancer for which screening was being done. However, the number of saved lives per unit of resources spent to screen was much lower than the number saved if some resources were refocused in other health issues. The costs were worth it from the afflicted individual's vantage but not from that of the population as a whole. Nor was the original screening regime worthwhile for the healthy individual. The risk of diminished health for an individual with a false positive screening result actually increases due to possible complications during follow-up biopsies (Schroder et al., 2009). Finally, the current screening regimes were not advantageous from the vantage of an individual whose life might have been saved by increased screening intensity for another deadly disease using reallocated funds. Hence, although this is seemingly counterintuitive, benefit and risk depend profoundly on vantage.

Understanding the vantage from which risk is framed is also critical for the management of environmental toxicants. The mercury in tainted fish eaten by Minamata victims constituted an unacceptable risk to the Japanese citizens; however, any extremely small increase in risk of autism from vaccination is trivial relative to risks from the diseases (diphtheria, tetanus, pertussis, or H1N1 influenza) for which vaccination is recommended. The benefit in lives saved to the entire population of children is also much higher than the dubious risk of autism to one's vaccinated child given that recent epidemiological evidence has confirmed that low-level mercury poisoning in children is not a cause of autism (Ng et al., 2007). Also, the risk from mercury to a population consuming modest amounts of seafood might be trivial but that to a subpopulation eating large amounts of seafood should warrant careful scrutiny. Likewise, the risk to a particularly vulnerable subpopulation such as unborn children exposed to mercury via food consumption by their mother should warrant much more scrutiny than that to a mature adult. Again, only the clarity associated with quantifying risk makes everything obvious from these different vantages. Without this decisions are merely based on insubstantial opinion and confused vantages.

## MEASURING EFFECT LEVELS

What is needed to estimate risk from environmental mercury? The most important mercury source for the general public is seafood so information is needed about mercury in seafood, seafood consumption rates for different groups of people, and the relationship between the amount of mercury to which a person is exposed and the likelihood of it manifesting some adverse effect. For humans this risk might be estimated from the vantage of the entire population (e.g., average US citizen), a sensitive subgroup (e.g., women of childbearing age), or a highly exposed subpopulation (e.g., fishermen or recent Vietnamese residents of the United States who consume more fish than the average citizen). Although human risk assessments require careful consideration of vantage, establishing the vantage for ecological risk estimation requires even more care because exposure might involve many other sources of mercury (i.e., via air, water, soil/sediment, or food chain) and different organisms that vary widely in terms of their mercury exposure route, uptake rate, and sensitivity.

## Human Harm

Existing exposure and effects data allow reasonable estimates of mercury risk. Initial efforts to set the limits for amount of mercury that can be ingested (expressed as oral reference doses or RfDs) drew on an accidental poisoning of Iraqis who unknowingly milled methylmercury-fungicide-treated seed into baking flour (Bakir et al., 1973). The limits for daily mercury ingestion in food (0.0001 mg/kg of body weight per day) were calculated after neurological testing of 81 children born to mothers who mistakenly consumed different amounts of this flour while pregnant (Bakir et al., 1973; Crump et al., 2000).*

Reluctance to treat these estimates as definitive was founded on the nature of the exposure. The Iraqi poisonings involved brief and intense exposures but those for which we commonly wish to estimate risk are exposures to low concentrations in seafood for long periods of time. Epidemiologists sought out communities that consumed high amounts of seafood with the intent of quantifying ingestion limits more definitively. Again, the vantage was harm to unborn children from a mother's exposure during pregnancy. Mother–child pairs from the Seychelles Islands in the Indian Ocean (Davidson et al., 1998; Crump et al., 2000; van Wijngaarden et al., 2006) and from the Faroe Islands in the North Atlantic (Grandjean et al., 1997) were meticulously monitored for mother exposure levels and tested for neurological impairment of their children. Inhabitants of the Seychelles consume fish that have mercury concentrations similar to those in the US market, but they consume 10 to 20 times more fish than most US citizens (Davidson et al., 1998). The Nordic inhabitants of the Faroe Islands consume large amounts of fish; however, they also consume pilot whale meat that contains high mercury concentrations in the range of 1.6 μg/g. Both of these epidemiological studies produced a more definitive and directly relevant limit for mercury ingestion (0.0001 mg/kg maternal body weight daily).† Ingestion rates above this limit might carry unacceptable risk to unborn children (Crump et al., 2000; van Wijngaarden et al., 2006).

The United States Environmental Protection Agency (US EPA) maintains an Integrated Risk Information System, or IRIS, that synthesizes these and other study results from the scientific literature. It can be accessed at this time from http://www.epa.gov/ncea/iris/index.html. Information for quantitatively estimating the limits for mercury exposure can be found by entering "Methylmercury (MeHg)" into the IRIS front page search tool. A summary document or full report on methylmercury is obtained by simply choosing one or the other on the Advanced Search Results from the methylmercury search. For example (Figure 4.2) the summary provides the daily oral reference dose (RfD) of 0.0001 mg of mercury/kg of person body weight per day.

* This corresponds to a maternal hair mercury concentration in the range of 10 to 20 μg/g (Grandjean et al., 1997), which is in the range of the current health criteria for pregnant women (Fujino, 1994). Hair mercury concentrations were as high as 674 μg/g for poisoned Iraqis, and at that level were linked to seizures, cerebral palsy, and other serious consequences to victims of acute poisoning (Davidson et al., 1998). Adult Japanese displaying neurological or neuropsychiatric disturbances notionally from chronic exposure to mercury in seafood had hair mercury concentrations as high as 37.4 μg/g, and averaging 7.9 μg/g in women and 11.6 μg/g in men (Fujino, 1994).

† This corresponds to maternal hair mercury concentrations of approximately 20 to 25 μg/g of hair.

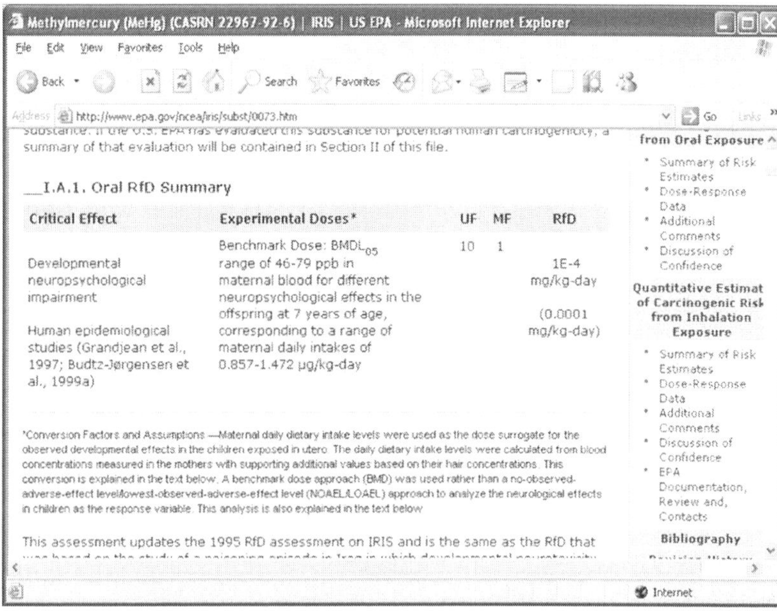


**FIGURE 4.2** Screen shot of US EPA's IRIS methylmercury summary for human oral ingestion.

This dose is based on the aforementioned epidemiological study of neuropsychological effects on seven-year-old children born of Faroe Islands mothers whose estimated daily dietary intakes during pregnancy ranged from 0.00086 mg to 0.00147 mg mercury/kg body weight (Budtz-Jørgensen et al., 1999; Grandjean et al., 1997). The mother's dose of concern was denoted the Benchmark Dose ($BMDL_{05}$), a dose with an excess risk of 5% for animals with abnormal levels. The dose ($BMDL_{05}$ = 0.00086 mg/kg-day) for this case was divided by a conservative uncertainty factor (UF) of 10 to produce the final reference dose of 0.000086, which is rounded to 0.0001 mg/kg-day. Risk of neurological harm to an unborn child is judged as potentially unacceptable if a mother eats more than that amount of mercury each day while pregnant.

## Ecological Harm

Risk to valued ecological entities, such as a charismatic species of fish hawk, is much more difficult to quantify although direct laboratory exposures of nonhuman species are possible whereas deliberate laboratory exposures of humans would be unethical. Regardless, there are so many animals and plants to consider that information is often inadequate to accurately estimate ecological risk for any particular species. Consequently, ecological risk assessments draw more from the original primary reports in the literature than from orderly compilations of information such as IRIS. The US EPA does have a data compilation tool called ECOTOX that is accessible from http://cfpub.epa.gov/ecotox/. A Quick Database Query specifying "Animals → Chloromethylmercury → Endpoint Reported → All Effect types → Publications from

1980 to 2009 → Viewable Table" will yield a table of effect/concentration data for a series of organisms. (Click on the Key Functions and then Query for Aquatic Biota to begin the search.) The resulting data will be a mixture of various measures of effect at specified mercury concentrations. For aquatic organisms, concentrations will be predominantly for methylmercury dissolved in water. This compromises many ecological risk assessments because most exposure of aquatic animals to methylmercury occurs from ingestion of contaminated food. You can do the search again and specify terrestrial species to get information on land organisms. The results will be biased toward bird species because they tend to be the focus of public attention during ecological risk assessments. Some results will be expressed as the median lethal concentration (LC50), median effect concentration (EC50), no observed effect level (NOEL), or lowest observed effect level (LOEL) values. The LC50 is the concentration calculated to kill half of the individuals exposed to that concentration, and the EC50 is the concentration that causes adverse effect (e.g., growth inhibition or reproductive impairment) to half of the test population. The LOEL is the lowest test treatment level in a toxicity test at which a statistically significant effect was noted and the NOEL is the next lowest test treatment level below the LOEL (Figure 4.3). Like the human health assessment community (Filipsson et al., 2003; Crump, 1984), ecological risk assessors are gradually coming to realize that LOEL or NOEL measures are compromised (Newman, 2008; Warne and van Dam, 2008) and should be replaced by metrics similar to benchmark doses.

But what exactly are these measures of effect? This question can be answered with a single example (Figure 4.3). This figure depicts data from a fictitious study quantifying mercury effect at different concentration or dose treatments. Five treatment groups of ducklings, including a control group, are fed different doses or concentrations of methylmercury for a set amount of time and then some indicator of neurological harm is measured for each of ten ducklings in each of the three replicates for each group. Some expression of the number of afflicted ducklings (mean and standard error) is plotted against the dose. To estimate the benchmark dose (i.e., BMD) for a predetermined effect level such as 5% of ducklings afflicted or 5% decrease in the neurological function, some regression model is fit to these data and a prediction made for the 5% level. Often the prediction is made in a conservative way by using some highest reasonable slope such as the statistical upper limit of the estimated model slope. The intent is to err on the side of safety in predictions of the dose (or concentration) at which that level of effect might be expected. The median lethal dose (LD50) or LC50 is also estimated from the same regression model. It is the concentration or dose predicted to kill 50% of the exposed ducklings. Sometimes, as in this example, the effect might not be lethal and an effective dose or concentration (ED50 or EC50) would be calculated.

Sometimes a model is not assumed and instead the effect noted in each treatment is tested statistically to see if it is different from the effect measured in the control treatment (e.g., using one-way analysis of variance and Dunnett's post hoc multiple comparison test). Effect levels in some treatments might be significantly different from the control (denoted in Figure 4.3 with an "S" next to the data point) while others might not be. The lowest treatment dose or concentration with a significant difference is called the LOEL and the next lower treatment down is called the NOEL.

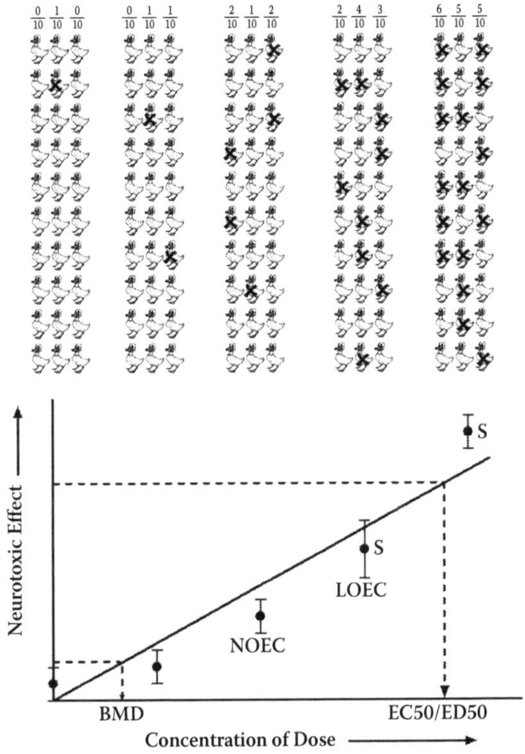


**FIGURE 4.3** Analysis of data from a fictitious experiment quantifying neurotoxicity in ducklings fed different doses or concentrations of methylmercury. Each treatment involved three replicates of ten ducklings exposed for a specific time period and then examined for neurological effect. For the sake of clarity, a straight line model is fitted to the mean values (± standard error) of the neurological effect response (lower panel), although sigmoid (S-shaped) models are often applied to these types of data. At the top of this figure is a depiction of the ten ducklings per replicate with the numbers adversely affected per ten ducklings shown, e.g., 0/10 1/10, 0/10, etc. With the neurological effect response for each treatment, it is feasible to determine the NOEL and LOEL using analysis of variance and Dunnett's test and compute the $BMDL_{05}$ and EC50/ED50 values using the regression model. (See text for more explanation.)

The LOEL and NOEL are thought to define the interval containing the threshold level needed to get an adverse effect. However, there are many difficulties with this approach that are discussed in publications such as Newman (2008) and Warne and van Dam (2008).

A good illustration of currently deficient information for doing risk assessments is the database for birds. Often, but not always, the species for which data are first generated are those easily reared and exposed in the laboratory. Accordingly, feeding trials for the mallard duck (*Anas platyrhynchos*) indicate that ducklings from hens fed 3 µg/g of mercury (as methylmercury) had abnormal avoidance behavior (Heinz,

1976). Ducklings with neurological indications of methylmercury poisoning were associated with eggs having 2 μg/g of mercury due to hen exposure through their feed (Heinz and Hoffman, 2003). Mallard hens receiving 0.5 μg/g as methylmercury in their feed laid fewer eggs and produced fewer ducklings than control hens, and their ducklings were hyper-responsive to alarm stimuli (Heinz, 1979).* These results were similar to those with young captive egrets (*Ardea albus*) fed 0.5 μg/g (as methylmercury) that displayed depressed general activity and prey hunting behavior relative to control egrets (Bouton et al., 1999). This 0.5 μg/g was also similar to those of egret prey species in the Florida Everglades. Surveys similar to the epidemiological studies described for humans are also done to determine concentrations resulting in effects to other species. For example, the survey of common loons (*Gavia immer*) by Meyer et al. (1998) suggested that egg mercury concentrations in lakes with lower reproductive success were similar to egg concentrations in experimentally dosed birds showing an adverse effect (Fimreite, 1971). Comprehensive surveys exist for blood, feather, and egg concentrations in field populations of this fish-eating bird (Evers et al., 1998; Meyer et al., 1998). Laboratory exposures of loons and examination of their chicks suggest mercury concentrations to which loons were exposed might decrease the immunological competency of the chicks (Kenow et al., 2007a).

## MEASURING EXPOSURE

The co-occurrence of mercury and an organism does not necessarily mean that that organism is exposed in a way that will cause harm. The form of mercury and the nature of the contact are extremely important in determining the realized exposure.

Mercury can be present in inorganic and organic forms that differ widely in their ability to enter an organism and cause harm. For example, the mercury† in silver-mercury amalgam dental fillings is only modestly available to enter the bloodstream (Gundacker et al., 2006), move into nervous tissues, and cause neurological harm. Mercury combined with selenium (i.e., HgSe) is also extremely unavailable. Indeed, toothed whales have evolved a detoxification mechanism that incorporates mercury into inert HgSe granules in tissues (Mackey et al., 2003). At the other extreme, contact with just a minuscule amount of the extremely bioavailable dimethylmercury can kill a person.‡ Dimethylmercury is very soluble in lipids such as the oils and fats in our cells. It evaporates readily and can be inhaled. Anyone in contact with dimethylmercury will quickly absorb it through the skin or lung surfaces, and the compound will move through the circulatory system to nervous tissue where it can cause lethal

* The mercury concentration of 0.5 μg/g in the feed can be converted to an ingestion rate as described later for humans. These mallard duck hens ate an average of 156 g of feed/kg body mass each day. So 0.5 μg/g × 156 g/kg-daily = 78 μg/kg-day or 0.078 mg/kg-day of mercury as methylmercury was the ingestion rate associated with these adverse effects. This also corresponded with approximate mercury concentrations of 10 μg/g of hen primary feathers and 0.8 μg/g wet weight in eggs (Heinz, 1979).

† This elemental mercury is in the form, $Hg^0$, that is a mercury atom with no charge. The other common form of inorganic mercury in the environment is divalent mercury, $Hg^{2+}$, which can bind to sediments, soils, and molecules in tissues.

‡ Mercury can combine with one or two methyl- groups to form either monomethylmercury ($HgCH_3$) or dimethylmercury ($H_3CHgCH_3$). Both are called methylmercury but the most common form in the environment is monomethylmercury.

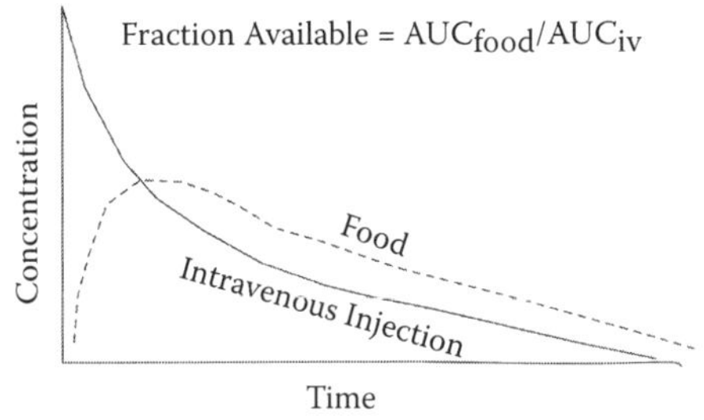


**FIGURE 4.4** An example of quantifying bioavailable mercury in food. A dose of mercury (e.g., 0.5 mg of mercury per kg body weight) is injected directly into the bloodstream and the decrease in blood mercury concentration measured over time (solid line). The area under the curve (AUC) is estimated with a model or the simple trapezoid method. The same dose is now administered in the food item of interest and the curve of blood mercury concentration monitored (dashed line). The AUC is estimated for this second curve. By definition the amount of mercury injected into the blood stream is 100% bioavailable so the quotient, $AUC_{food}/AUC_{iv}$, is an estimate of the fraction of the ingested dose available to enter the blood stream.

damage. Sadly this was the case in 1996 when brief contact of a toxicology professor's gloved hand with a drop of spilled dimethylmercury led to her death within ten months (Nierenberg et al., 1998).* Intermediate between the bioavailabilities of the mercury in HgSe granules and dimethylmercury is that of monomethylmercury. Monomethylmercury, along with inorganic forms of mercury, comprises most of the mercury found in seafood (Mason et al., 2006). It is bound to sulfur-rich molecules in tissues such as proteins or glutathione (Harris et al., 2003) and is thought to be relatively bioavailable after ingestion.

But descriptors such as "sparingly," "modestly," "extremely," or "relatively" bioavailable are too vague to help in risk assessments; therefore, quantitative measures of bioavailability were created. Usually toxicant bioavailability after ingestion, inhalation, or absorption across the skin is quantified as that proportion of the mercury that gets into the bloodstream (Figure 4.4). This fraction is important to quantify, but a risk assessor might be forced to assume that 100% of the mercury in a source is bioavailable if no bioavailability estimate exists.

Other qualities of the contact with the mercury also need to be understood in order to estimate risk. This is easiest to illustrate with the equation often used for seafood ingestion by regulatory agencies,

$$\text{Intake (mg/kg per day)} = \frac{(\text{CF})(\text{IR})(\text{FI})(\text{EF})(\text{ED})}{(\text{BW})(\text{AT})}$$

where CF = concentration in the food (mg/kg) that might be modified with the bioavailability fraction to the concentration of bioavailable mercury if only a portion of

* Peak hair mercury concentrations (approximately 1100 μg/g) were reached in a little more than a month after the accidental exposure.

the mercury in a food is bioavailable; IR = ingestion rate (kg/meal), e.g., 0.284 kg/meal (upper 95% confidence level for fish) consumption; FI = fraction ingested from the contaminated source (no units to this fraction); EF = exposure frequency (meals/year); ED = exposure duration (years), e.g., 0.75 for a human gestation period; BW = body weight or mass (kg), e.g., 70 kg for an adult; and AT = averaging time (days), e.g., ED × 365 days.

Information needed to estimate intake is generated carefully for the general population or for a particular subpopulation of concern. Conservative values might be entered into the equation if information for the general population is compiled. As an example, the use of a value at the upper 95% confidence limit for the mean seafood ingestion rate for the public can embed another safety factor in the risk assessment process.

## Human Exposure

Many research programs generate the information needed to estimate human exposure through fish consumption. Most often the emphasis is on maternal exposure during pregnancy so ED is 0.75 years and AT is 365 (0.75), or 274 days in the above equation. Remaining undefined after this vantage is established are CF, BW, and factors related to consumption habits (IR, FI, EF). These pieces of information are sometimes applied in the form of means or medians; however, increasingly often they are applied as entire distributions using computer simulation methodologies (e.g., Monte Carlo simulation).

Body weight of women is the most obvious item in the equation with which to show the advantage of considering information as distributions. A generic body weight (BW) of 70 kg is often not accurate enough because, as in the case of North American women of child-bearing age, body weights are usually lower (mean: 60.6 kg) and can vary widely (standard deviation: 11.9 kg) (US EPA, 1997). Whether the weight is selected from the center or some other position of the distribution for the population will depend on the consequences of being wrong and the specific group for which risk is being estimated. To be conservative an assessor might pick a smaller woman's weight, such as that associated with a woman in the lower 10% of the population weight distribution.

Mercury concentrations in the ingested food (CF) also vary within and among seafood species. Considerable effort is exerted to survey mercury in potential seafood species. Surveys might emphasize the countrywide seafood market (e.g., Sunderland, 2007), a particular regional market (e.g., Hammerschmidt and Fitzgerald, 2006; Mason et al., 2006; Cheung et al., 2008; Monson, 2009; Peterson et al., 2007), the species eaten in a contaminated region (e.g., Amazon fishes studied by Regine et al., 2006; and Wujiang River fishes surveyed by Li et al., 2009), species high in mercury (e.g., tuna studied by Krapiel et al., 2003), or a species consumed by a particular subpopulation (e.g., small cetaceans eaten by Japanese as noted in Endo et al., 2005). The concentration used in the equation might be a composite of concentrations in different items eaten by the population of interest. A careful assessment might require using distributions of mercury concentrations in many different seafood items (Tang et al., 2009).

Careful surveys of the general population or specific populations (e.g., secondary school students of Hong Kong (Tang et al., 2009)) are used to establish information about ingestion rate (IR), fraction taken from the source of concern (FI), and the exposure frequency (EF). The detail and breadth required in a particular survey depends on the context. An initial estimate of exposure might be generated with information about the average US woman (i.e., 1.8 g/day) (Holloman and Newman, 2010). Ingestion rates for specific seafood items may be used if there are material differences in mercury concentrations in or preference for certain seafood items (US EPA, 2003; Tang et al., 2009). However, results based on generic information could be misguiding in some situations. For example, the Japanese noted as having neurological dysfunction (footnote on page 56, this volume) had ingested seafood at a rate of 333.6 g/day (Fujino, 1994). African-American women in Newport News, Virginia, eat 147.8 g/day of seafood (Holloman and Newman, 2010). Relative to EF, Swedish women surveyed by Bjornberg et al. (2005) ate fish 1.6 to 19 times weekly, but women living along the St. Lawrence River ate fish less frequently (0 to 7.9 times weekly) (Morrissette et al., 2004). Similarly, secondary school children of Hong Kong varied materially in their fish consumption rate with estimated dietary methylmercury exposures ranging from 0.4 to 0.5 μg/kg body weight-weekly (for average consumers) to 1.2 to 1.4 μg/kg body weight-weekly (for high consumers) (Tang et al., 2009). In such cases the vantage might be refined so as to better assess the health hazard to each subpopulation of concern.

Less detailed exposure assessments are done, but they provide less insight for fully understanding the exposure. This lack of detail can make it difficult to determine what changes in behavior might reduce risk. An important, previously mentioned example was exposure expressed as mercury concentration in hair (Table 4.1). Maternal blood, umbilical cord blood, fingernails, and other samples might be considered as surrogates for exposure concentrations and correlated with effects instead of doing the exhaustive data gathering needed to apply the above ingestion rate equation (e.g., Fok et al., 2007).

## Ecological Exposure

Exposures of the many valued ecological entities by a variety of pathways are extremely difficult to calculate accurately except in the most fortuitous of situations. The epidemiological surveys, similar to those done to define human exposures, are important, but laboratory experiments are also prominent for nonhuman species' exposures. Because so many species must be considered, information for a surrogate species is commonly employed. For example, exposure information for a fish hawk might be collected with the intention of comparing it to exposure-effect information generated in the laboratory for mallard ducks; marine minks might be used as surrogates for dolphins and seals (Hung et al., 2007). Often an ecological risk assessor is forced into the awkward logic of "it doesn't walk or quack like a duck but, for purposes of this assessment, it's a duck." The mercury in food eaten by mallard ducks could reasonably be assumed to have a different bioavailability when compared with that in the fish eaten by a fish hawk. Obviously, the sound work with mallard ducks is directly relevant to this and similar important duck species but questionable in other

**TABLE 4.1**
**Total Mercury Concentration in Human Hair as a Measure of Exposure**

| μg Hg/g Hair | Study Subject | Reference |
|---|---|---|
| 0.09 | Pregnant women living along St. Lawrence River, GM | Morrissette et al. (2004) |
| 0.12 | US children ages 1–5 years (1999–2000), GM | McDowell et al. (2004) |
| 0.20 | US women ages 16–49 years (1999–2000), GM | McDowell et al. (2004) |
| 0.38 | US women as above but frequent fish consumers, GM | McDowell et al. (2004) |
| 0.38 | Vegetarian (vegan) men in Hong Kong, Mean | Dickman et al. (1998) |
| 0.7 | Swedish women, Med | Bjornberg et al. (2005) |
| 1.9 | Coastal Brazilian population, Mean | Nilson et al. (2001) |
| 1.6 | Hong Kong women, Mean | Dickmam et al. (1998) |
| 1.6 | UK individuals eating fish 1 to 4 times monthly | Airey (1983) |
| 2.4 | US individuals eating fish 1 to 4 times monthly | Airey (1983) |
| 2.5 | Australian eating fish 1 to 4 times monthly | Airey (1983) |
| 2.6 | Fertile Hong Kong men, Mean | Dickman et al. (1998) |
| 4.5 | Subfertile Hong Kong men, Mean | Dickman et al. (1998) |
| 4.3 | Faroese study, new mothers, GM | Grandjean et al. (1997) |
| 6.8 | Maternal hair from Seychelles study | Davidson et al. (1998) |
| 7.9 | Men, Katsurajima Island, Japan (1974–79), Mean | Fijino (1994) |
| 10 | Maternal hair health upper limit, Faroe study | Grandjean et al. (1997) |
| 11.6 | Women, Katsurajima Island, Japan (1974–79), Mean | Fijino (1994) |
| 25 | Estimated to correspond with current ingestion limit | Crump et al. (2000) |
| 0.9-28.5 | Range in Peruvian fishing village of Mancora | Marsh (1995) |
| 21.5-33.9 | Range in Wuchuan mining area of Guizhou, China | Li et al. (2008) |
| 37.4 | Highest value, Katsurajima Island, Japan (1974–79) | Fijino (1994) |
| 100–191 | Asymptomatic Minamata residents, 1960 survey | Smith and Smith (1975) |
| 0.16–199 | Range in Wujiazhan, China (mean = 3.41) | Zhang and Wang (2006) |
| 96.8–705 | Symptomatic Minamata residents, 1960 survey | Smith and Smith (1975) |
| 674 | Iraqis poisoned by methylmercury-treated grain | Davidson et al. (1998) |
| 1100 | Maximum during acute dimethylmercury poisoning | Nierenberg et al. (1998) |

*Note:* Maternal hair mercury concentration in the range of 10 to 25 μg/g is indicative of possible harm to unborn children based on compilations of these types of postexposure data.
GM = geometric mean, Mean = arithmetic mean, Med = median

instances. Much effort is being spent at this time to produce information needed to reduce the inaccuracy of such compromised assessments.

However, information directly useful for relevant species is increasingly being produced for assessing mercury exposure. As an example, the information base for the fish-eating common loon, *Gavia immer*, is growing rapidly. Surveys have been published that describe the geographic distribution of mercury measured in feathers and blood of North American loon populations (Evers et al., 1998). The exposure measures of feather and blood mercury have been correlated in field surveys with adult loon reproduction and survival (Meyer et al., 1998). Results from these field studies have been integrated with those from laboratory exposures of loon chicks to suggest linkage between mercury exposure and adverse effects (Kenow et al., 2007a, 2007b). Another example is exposure information (often hair concentrations) of wild mammals such as mink (Moore et al., 1999), arctic fox (Fuglei et al., 2007), and polar bear (Dietz et al., 2006).

## CONCLUSIONS

Selecting an appropriate vantage and then quantifying exposure and an associated effect level are essential to understanding enough to make a reasonable judgment about risk from environmental mercury. Dependence on unreliable opinions and widely divergent information is the only other option in the absence of these essential steps. Fortunately enough information is slowly emerging to make wise decisions for environmental risk assessment of mercury. A well-informed quantitative risk assessment not only reveals the current hazard of mercury to human populations and ecosystems of concern, but also provides essential information such as recommended safe food items and their allowable intake rates for effective risk management of environmental mercury. This short chapter sketches out the associated information and provides the general quantitative approach. The interested reader is urged to explore the references cited below for more information.

## REFERENCES

Adams, John. 1995. *Risk.* London: UCL Press Limited.

Airey, D. 1983. Total mercury concentrations in human hair from 13 countries in relation to fish consumption and location. *Science of the Total Environment,* 31:157–180.

Andriole, Gerald L., E. David Crawford, Robert L. Grubb III, Sandra S. Buys, David Chia, Timothy R. Church, Mona N. Fouad, Edward P. Gelmann, Paul A. Kvale, Douglas J. Reding, Joel L. Weissfeld, Lance A. Yokochi, Barbara O'Brien, Jonathan D. Clapp, Joshua M. Rathmell, Thomas L. Riley, Richard B. Hayes, Barnett S. Kramer, Grant Izmirlian, Anthony B. Miller, Paul F. Pinsky, Philip C. Prorok, John K. Gohangan, and Christine D. Berg. 2009. Mortality results from a randomized prostate-cancer screening trial. *The New England Journal of Medicine,* 360:1310–1319.

Bakir, F., S.F. Damluji, L. Amin-Zaki, M. Murtadha, A. Khalidi, N.Y. Al-Rawi, S. Tikriti, H.I. Dhahir, T.W. Clarkson, J.C. Smith, and R.A. Doherty. 1973. Methylmercury poisoning in Iraq. *Science,* 181:230–241.

Beamish, Thomas D. 2001. Environmental hazard and institutional betrayal. *Organization & Environment,* 14:5–33.

Bjornberg, Karolin A., Marie Vahter, Kiersten P. Grawe, and Marika Berglund. 2005. Methyl mercury exposure in Swedish women with high fish consumption. *Science of the Total Environment,* 341:45–52.

Bouton, Shannon N., Peter C. Frederick, Marilyn G. Spalding, and Heather McGill. 1999. Effects of chronic, low concentrations of dietary methylmercury on the behavior of juvenile great egrets. *Environmental Toxicology and Chemistry,* 18:1934–1939.

Budtz-Jørgensen, Esben, Neils Keiding, Phillipe Grandjean, Roberta F. White, and Pal Weihe. 1999. Methylmercury neurotoxicity independent of PCB exposure. *Environmental Health Perspectives,* 107:A236–237.

Cheung, K.C., H.M. Leung, and M.H. Wong. 2008. Metal concentrations of common freshwater and marine fish from the Pearl River Delta, South China. *Archives of Environmental Contamination and Toxicology,* 54:705–715.

Crump, Kenny S. 1984. A new method for determining allowable daily intakes. *Fundamental and Applied Toxicology*, 4:854–871.

Crump, Kenny S., Cynthia van Landingham, Conrad Shamlaye, Christopher Cox, Phillip W. Davidson, Gary J. Meyers, and Thomas W. Clarkson. 2000. Benchmark concentrations for methylmercury obtained from the Seychelles Child Development Study. *Enviromental Health Perspectives,* 108:257–263.

Davidson, Philip W., Gary J. Myers, Christopher Cox, Catherine Axtell, Conrad Shamlaye, Jean Sloane-Reeves, Elsa Cernichiari, Larry Needham, Anna Choi, Yining Wang, Maths Berlin, and Thomas W. Clarkson. 1998. Effects of prenatal and postnatal methylmercury exposure from fish consumption on neurodevelopment. *JAMA,* 280:701–707.

Dickman, Mike D., Clement K.M. Leung, and Milton K.H. Leong. 1998. Hong Kong male subfertility links to mercury in human hair and fish. *Science of the Total Enviroment,* 214:165–174.

Dietz, Rune, Frank Riget, Erik W. Born, Christian Sonne, Philippe Grandjean, Maja Kirkegaard, Morten T. Olsen, Gert Asmund, Aristeo Renzoni, Hans Baagoe, and Christian Andreasen. 2006. Trends in mercury in hair of Greenland polar bears (*Ursus maritimus*) during 1892–2001. *Environmental Science & Technology,* 40:1120–1125.

Endo, Tetsuya, Koichi Haraguchi, Yohei Hotta, Yohsuke Hisamichi, Shane Lavery, Merel L. Dalebout, and C. Scott Baker. 2005. Total mercury, methyl mercury, and selenium levels in the red meat of small cetaceans sold for human consumption in Japan. *Environmental Science & Technology,* 39:5703–5308.

Evers, David C., Joseph D. Kaplan, Michael W. Meyer, Peter S. Reaman, W. Emmett Braselton, Andrew Major, Neil Burgess, and Anton M. Scheuhammer. 1998. Geographic trend in mercury measured in common loon feathers and blood. *Environmental Toxicology and Chemistry,* 17:173–183.

Filipsson, Agneta F., Salomon Sand, John Nilsson, and Katarina Victorin. 2003. The benchmark dose method—Review of available models, and recommendations for application in health risk assessment. *Critical Reviews in Toxicology,* 33:505–542.

Fimreite, Norvald. 1971. Effects of methylmercury on ring-necked pheasants. *Canadian Wildlife Service Occasional Paper,* 9:39.

Fok, Tai F., Hugh S. Lam, Pak C. Ng, Alexander S.K. Yip, Ngai C. Sin, Iris H.S. Chan, Goldie J.S. Gu, Hung K. So, Eric M.C. Wong, and Christopher W.K. Lam. 2007. Fetal methylmercury exposure as measured by cord blood mercury concentrations in a mother-infant cohort in Hong Kong. *Environment International,* 33:84–92.

Fuglei, Eva, Jan O. Bustnes, Haakon Hop, Torill Mork, Helen Bjornfoth, and Bert van Bavel. 2007. Environmental contaminants in arctic foxes (*Alpex lagopus*) in Svalbard: Relationships with feeding ecology and body condition. *Environmental Pollution,* 146:128–138.

Fujino, Tadashi 1994. Clinical and epidemiological studies on chronic Minamata disease. Part I: Study on Katsurajima Island. *Kumamoto Medical Journal,* 44:139–155.

Gigerenzer, Gerd. 2002. *Calculated Risk.* New York: Simon & Schuster.

Grandjean, Philippe, Pal Weihe, Roberta F. White, Frodi Debes, Shunichi Araki, Kazuhito Yokoyama, Katsuyuki Murata, Nicolina Sørensen, Rasmus Dahl, and Poul J. Jørgensen. 1997. Cognitive deficit in 7-year-old children with prenatal exposure to methylmercury. *Neurotoxicology and Teratology,* 20:1–12.

Gundacker, Claudia, Günter Komarnicki, Bettina Zödl, Christian Forster, Ernst Schuster, and Karl Wittmann. 2006. Whole blood mercury and selenium concentrations in a selected Austrian population: Does gender matter? *Science of the Total Environment,* 372:76–86.

Hammerschmidt, Chad R., and William F. Fitzgerald. 2006. Bioaccumulation and trophic transfer of methylmercury in Long Island Sound. *Archives of Environmental Contamination and Toxicology,* 51:416–424.

Harris, Hugh H., Ingrid J. Pickering, and Graham N. George. 2003. The chemical form of mercury in fish. *Science,* 301:1203.

Heinz, Gary H. 1976. Methylmercury: Second-year feeding effects on mallard reproduction and duckling behavior. *Journal of Wildlife Management,* 40:82–90.

Heinz, Gary H. 1979. Methylmercury: Reproductive and behavioral effects on three generations of mallard ducks. *Journal of Wildlife Management,* 43:394–401.

Heinz, Gary H., and David J. Hoffman. 2003. Embryonic thresholds of mercury: Estimates from individual mallard eggs. *Archives of Environmental Contamination and Toxicology,* 44:257–264.

Holloman, Erica L., and Michael C. Newman. 2010. A community-based assessment of seafood consumption along the lower James River, Virginia, USA: Potential sources of mercury exposure. *Environmental Research,* doi:10.1016/j.envres.2010.01.002.

Hung, Craig L.H., Ridge K.F. Lau, James C.W. Lam, Thomas A. Jefferson, Samuel K. Hung, Michael H.W. Lam, and Paul K.S. Lam. 2007. Risk assessment of trace elements in the stomach contents of Indo-Pacific humpback dolphins and finless porpoises in Hong Kong waters. *Chemosphere,* 66:1175–1182.

Kasperon, Roger E., and Jeanne X. Kasperson. 1996. The social amplification and attenuation of risk. *Annals of the American Academy of Political and Social Science,* 545: 95–105.

Kenow, Kevin P., Keith A. Grasman, Randy K. Hines, Michael W. Meyer, Annette Gendron-Fitzpatrick, Marilyn G. Spalding, and Brian R. Gray. 2007a. Effects of methylmercury exposure on the immune function of juvenile common loons (*Gavia immer*). *Environmental Toxicology and Chemistry,* 26:1460–1469.

Kenow, Kevin P., Michael W. Meyer, Randy K. Hines, and William H. Karasov. 2007b. Distribution and accumulation of mercury in tissues of captive-reared common loon (*Gavia immer*) chicks. *Environmental Toxicology and Chemistry,* 26:1047–1055.

Kraepiel, Anne M., Klaus Keller, Henry B. Chin, Elizabeth C. Malcolm, and François M.M. Morel. 2003. Sources and variations of mercury in tuna. *Environmental Science & Technology,* 37:5551–5558.

Li, Ping, Xinbin Feng, Guangle Qiu, Lihai Shang, and Shaofeng Wang. 2008. Mercury exposure in the population from Wuchuan mercury mining area, Guizhou, China. *Science of the Total Environment,* 395:72–79.

Li, Sixin, Lianfeng Zhou, Hongjun Wang, Youguang Liang, Juxiang Hu, and Jianbo Chang. 2009. Feeding habits and habitat preferences affecting mercury bioaccumulation in 37 subtropical fish species from Wujiang River, China. *Ecotoxicology,* 18:204–210.

Locke, John. 1690, reprinted 1959. *An Essay Concerning Human Understanding*. New York: Collated by Alexander C. Fraser and reprinted by Dover Publications Inc.

Mackey, Elizabeth A., Rabia D. Oflaz, Matthew S. Epstein, B. Buehler, Barbara .J. Porter, Teri Rowles, S.A. Wise, and Paul R. Becker. 2003. Elemental composition of liver and kidney tissues of rough-toothed dolphin (*Steno bredanensis*). *Archives of Environmental Contamination and Toxicology,* 44:523–532.

Marsh, D.O., M.D. Turner, J.C. Smith, P. Allen, and N. Richdale. 1995. Fetal methylmercury study in a Peruvian fish-eating population. *Neurotoxicology,* 16:717–726.

Marshall, Eliot. 2010. Brawling over mammography. *Science,* 327:936–938.

Mason, Robert P., Deborah Heyes, and Auja Sveinsdottir. 2006. Methylmercury concentrations in fish from tidal waters of the Chesapeake Bay. *Archives of Environmental Contamination and Toxicology,* 51:425–437.

McDowell, Margaret A., Charles F. Dillon, John Osterloh, P. Michael Bolger, Edo Pellizzari, Reshan Fernando, Ruben Montes de Oca, Susan E. Schober, Thomas Sinks, Robert L. Jones, and Kathryn R. Mahaffey. 2004. Hair mercury levels in U.S. children and women of childbearing age: Reference range data from NHANES 1999–2000. *Environmental Health Perspectives,* 112:1165–1171.

McGinnis, J. Michael, and William H. Foege. 1993. Actual causes of death in the United States. *JAMA,* 270:2207–2212.

Meyer, Michael W., David C. Evers, Jerry J. Hartigan, and Paul S. Rasmussen. 1998. Patterns of common loon (*Gavia immer*) mercury exposure, reproduction, and survival in Wisconsin, USA. *Environmental Toxicology and Chemistry,* 17:184–190.

Monson, Bruce A. 2009. Trend reversal of mercury concentrations in piscivorous fish from Minnesota lakes: 1982–2006. *Environmental Science & Technology,* 43:1750–1755.

Moore, Dwayne R.J., Bradley E. Sample, Glenn W. Suter, Benjamin R. Parkhurst, and R. Scott Teed. 1999. A probabilistic risk assessment of the effects of methylmercury and PCBs on mink and kingfishers along the East Fork Popular Creek, Oak Ridge, Tennessee, USA. *Environmental Toxicology and Chemistry,* 18:2941–2953.

Morrissette, Joelle, Larissa Takser, Genevieve St. Amour, Audrey Smargiassi, Julie Lafond, and Donna Mergler. 2004. Temporal variation of blood and hair mercury levels in pregnancy in relation to fish consumption history in a population living along the St. Lawrence River. *Environmental Research,* 95:363–374.

Newman, Michael C. 2008. "What exactly are you inferring?" A closer look at hypothesis testing. *Environmental Toxicology and Chemistry,* 27:1013–1019.

Ng, Daniel Kwok-Keung, Chung-Hong Chan, Man-Ting Soo, and Robert Shing-Yan Lee. 2007. Low-level chronic mercury exposure in children and adolescents: Meta-analysis. *Pediatrics International,* 49:80–87.

Nierenberg, David W., Richard E. Nordgren, Morris B. Chang, Richard W. Siegler, Michael B. Blayney, Fred Hochberg, Taft Y. Toribara, Elsa Cernichiari, and Thomas Clarkson. 1998. Delayed cerebellar disease and death after accidental exposure to dimethylmercury. *New England Journal of Medicine,* 338:1672–1676.

Nilson Jr., Sant'Anna, Monica Costa, and Hirokatsu Akagi. 2001. Total and methylmercury levels of a coastal human population and of fish from the Brazilian Northeast. *Environmental Science and Pollution Research,* 8:1–5.

Peterson, Spencer A., John Van Sickle, Alan T. Herlihy, and Robert M. Hughes. 2007. Mercury concentration in fish from streams and rivers throughout the western United States. *Environmental Science & Technology,* 41:58–65.

Petrie, Keith J., Elizabeth A. Broadbent, Nadine Kley, Dipl Psych, Rona Moss-Morris, Rob Horne, and Winfried Rief. 2005. Worries about modernity predict symptom complaints after environmental pesticide spraying. *Psychosomatic Medicine,* 67:778–782.

Piattelli-Palmarini, Massimo. 1994. *Inevitable Illusions.* New York: John Wiley & Sons.

Popper, Karl R. 1972. *Objective Knowledge. An Evolutionary Approach.* Oxford: Clarendon Press.

Regine, Maury-Bratchet, Dirrieu Gilles, Dominique Yannick, and Boudou Alain. 2006. Mercury distribution in fish organs and food regimes: Significant relationships from twelve species collected in French Guiana (Amazonian basin). *Science of the Total Environment,* 368:262–270.

Richerson, Peter J., and Robert Boyd. 2005. *Not by Genes Alone.* Chicago: University of Chicago Press.

Schröder, Fritz H., Jonas Hugosson, Monique J. Roobol, Teuvo L.J. Tammela, Stefano Ciatto, Vera Nelen, Maciej Kwiatkowski, Marcos Lujan, Hans Lilja, Marco Zappa, Louis J. Denis, Franz Recker, Antonio Berenguer, Liisa Määttänen, Chris H. Bangma, Gunnar Aus, Arnauld Villers, Xavier Rebillard, Theodorus van der Kwast, Bert G. Blijenberg, Sue M. Moss, Harry J. de Koning, and Anssi Auvinen. 2009. Screening and prostate-cancer mortality in a randomized European study. *New England Journal of Medicine,* 306:1320–1328.

Smith, W. Eugene, and Aileen M. Smith. 1975. *Minamata.* New York: Holt, Rinehart and Winston.

Sunderland, Elsie M. 2007. Mercury exposure from domestic and imported estuarine and marine fish in the U.S. seafood market. *Environmental Health Perspectives,* 115:235–242.

Tang, Anna Shiu Ping, Ka Ping Kwong, Stephen Wai Cheung Chung, Yuk Yin Ho, and Ying Xiao. 2009. Dietary exposure of Hong Kong secondary school students to total mercury and methylmercury from fish intake. *Food Additives & Contaminants Part B,* 2:8–14.

US EPA. 1997. *Exposure Factors Handbook (Final Report) 1997, EPA/600/P-95/002F a-c.* Washington: US Environmental Protection Agency.

US EPA. 2003. *Estimated Per Capita Fish Consumption in the United States. August 2002, EPA/821/C/02/003.* Washington: US Environmental Protection Agency,

Van Wijngaarden, Edwin, Christopher Beck, Conrad F. Shamlaye, Elsa Cernichiari, Philip W. Davidson, Gary J. Myers, and Thomas W. Clarkson. 2006. Benchmark concentrations for methyl mercury obtained from the 9-year follow-up of the Seychelles Child Development Study. *Neurotoxicology,* 27:702–709.

Vernimmen, Pierre, Pascal Quiry, Yann Le Fur, Maurizio Dallocchio, and Antonio Salvi. 2005. *Corporate Finance. Theory and Practice.* New York: Wiley & Sons.

Warne, Michael St. J., and Rick van Dam. 2008. NOEC and LOEC data should no longer be generated or used. *Australasian Journal of Ecotoxicology,* 14:1–5.

Winters, Winnie, Stephan Devriese, Ilse Van Diest, Benoit Nemery, Hendrik Veulemans, Paul Eelen, Karel Van de Woestijne, and Omer Van den Bergh. 2003. Media warnings about environmental pollution facilitate the acquisition of symptoms in response to chemical substances. *Psychosomatic Medicine,* 65:332–338.

Zhang, Lei and Qichao Wang. 2006. Preliminary study on health risk from mercury exposure to residents of Wujiazhan town on the Di'er Songhua River, Northeast China. *Environmental Geochemistry,* 28: 67–71.

# 5 Is Tuna Safe? A Sociological Analysis of Federal Fish Advisories

*Kelly Joyce*

## CONTENTS



> Really, it's simple: Government guidelines regarding tuna consumption are ONLY for pregnant women or women who may be trying to become pregnant. For the rest of consumers and the general public the best advice about tuna consumption is: Eat more of it!
>
> **—National Fisheries Institute, 2010**

## INTRODUCTION

In 2004 the Food and Drug Administration (FDA) and the Environmental Protection Agency (EPA) issued a joint fish consumption advisory (see Appendix I, Figures 5.1, 5.2). Despite the claim in the above quotation, the advisory addressed pregnant women, nursing mothers, women who may become pregnant, and young children. Due to increasing evidence about the effects of methylmercury on the developing brain, these groups were advised to not eat shark, swordfish, king mackerel, and tilefish; eat up to 12 ounces a week of a variety of fish and shellfish that are lower in mercury; and check local advisories about the safety of fish caught by family and friends in local lakes, rivers, and coastal areas.

Yet the following caveat appeared in the advisory. After noting that women and young children should eat up to 12 ounces of fish and shellfish low in mercury per week, the advisory explained, "Five of the most commonly eaten fish that are low in mercury are shrimp, canned light tuna, salmon, pollock, and catfish. Another commonly eaten fish, albacore ("white") tuna has more mercury than canned light tuna. So, when choosing your two meals of fish and shellfish, you may eat up to 6 ounces (one average meal) of albacore tuna per week."

FDA and EPA's recommendation was perplexing. Why did albacore tuna get its own special recommendation? What about other kinds of tuna such as bluefin, bigeye, or yellowfin? Was only the developing nervous system at risk? What about men, women who were not planning on getting pregnant, and older people? Could these groups eat as much fish as they would like without risk, despite the toxic effects of long-term exposure to mercury? And, finally, why was the warning only posted on the agencies' websites? Shouldn't it be displayed in grocery stores (where people might actually purchase fish)?

This chapter analyzes the assumptions in the production and distribution of the federal 2004 fish advisory. Such an analysis shows how FDA and EPA fish consumption recommendations developed from interactions between federal agencies, tuna industries, news media, and environmental awareness organizations. A sociological approach highlights how a confluence of social actors and events turns an issue into a social problem—one that is publicly recognized, debated, and addressed through regulation and policy (Mills, 1959). Federal and state agencies, academic-affiliated research scientists, citizen groups, journalists, and industries (and, at times, their contentious exchanges) all helped to define "eating contaminated fish" as a social problem. The process of recognizing—as a society—the risks associated with consuming methylmercury in tuna and other fish has not been smooth or consensus-based, nor has it resulted in the creation of a uniform knowledge. How federal agencies, industries, and environmental organizations understand the issue varies, and these varied understandings are ongoing and incomplete, making awareness of the adverse human health effects of consuming contaminated fish a social problem that is both contested and in the making.

In this chapter, I focus in particular on FDA and EPA. I show how these agencies have different forms of expertise and institutional cultures that in turn shape how each understands the risks associated with methylmercury consumption. FDA draws upon medical forms of expertise that emphasize clinical care, has a regulatory past that still shapes the agency's organizational structure, and primarily relies on industries to provide data about possible technological or pharmacological risks. In contrast, EPA emphasizes toxicology, risk assessment, and environmental epidemiology, conducts research in house as well as reviews data from industry and university scientists, and has a robust peer review process. These different forms of scientific expertise and institutional practices help give rise to two divergent ways of framing the dangers associated with fish consumption. As this chapter shows, FDA and EPA collaborated on the 2004 fish advisory, but differing constituencies, missions, and forms of expertise may cause the alliance to break apart.

## RESEARCH METHODOLOGY

To investigate FDA and EPA activities in relation to methylmercury exposure, I used a multimethod approach that included analysis of archival data, medical literature, media accounts, and website content. EPA and FDA reports, announcements published in the Federal Register, FDA consumer articles, and FDA meeting transcripts and minutes all provided primary documents for investigation. The Federal Register is the federal government's official journal (available in hard copy and online) that contains most routine publications and public notices of government agencies. As federal agencies, meeting minutes and transcripts and reports are also publicly available.

In addition to studying primary materials, I conducted a targeted literature review of the peer-reviewed epidemiological journal articles on methylmercury and human consumption of fish. I further researched news media representations of methylmercury in fish via the search engine LexisNexis Academic. I collected and analyzed publications, letters posted to the Federal Register, and website content presented by US environmental and industry organizations. Finally, I interviewed selected scientists to better understand the debates about and events related to FDA and EPA involvement in risk assessment and methylmercury consumption.

The data gathered was analyzed to better understand how FDA, EPA, key environmental organizations, and seafood industries approached and understood methylmercury exposure and consumption, and how these ideas changed over time. An inductive analytical approach was used to investigate each organization's strategies and practices. This approach provided insight into how each organization both understood and communicated the risks and benefits associated with tuna and other fish consumption to particular audiences.

## HOW DOES METHYLMERCURY GET IN FISH?

Mercury, a naturally occurring element that can exist in the air, water, or earth, is found in various forms including elemental mercury, inorganic mercury compounds, and organic mercury compounds such as methylmercury (EPA, 2010). Mercury enters the environment from both natural and human sources. Forest fires and volcanic eruptions, for example, can release mercury into the air and mercury can leach from rocks into waterways. Anthropogenic, or human, sources range from direct discharge into waterways by processing plants or mining activities to the improper disposal of products that contain mercury (e.g. thermometers, fluorescent lightbulbs, dry cell batteries, dental amalgams, car parts) into landfills. These products then leach mercury into the soil and water (see Chapter 3, Figure 3.1).

Coal-burning power plants are the largest human-caused source of mercury emissions into the air in the United States, accounting for more than 40% of all domestic human-caused mercury emissions (EPA, 2010). Mercury is found in many rocks including coal. When coal is burned (as it is in coal-burning power plants), mercury is released into the air.

The form mercury takes in the air is elemental mercury vapor ($Hg^0$). Elemental mercury vapor can undergo a photochemical oxidation to become divalent mercury ($Hg^{2+}$), which then combines with water vapor and settles into waterways and soil.

Mercury vapor moves with global air currents to settle in countries all over the world. EPA (2010) estimates that about one-quarter of US emissions from coal-burning power plants are deposited within the contiguous United States, and the remainder enters the global cycle. In the same way, mercury released into the air from sources outside the US—coal plants in China, for example—travels across the globe on air currents and settles in waterways everywhere, including in the US.

Scientists are still researching how and why mercury is converted into methylmercury and then accumulated in fish and shellfish. They are unsure how and where mercury is methylated in the ocean (Sunderland et al., 2009). In some aquatic environments such as lakes and rivers mercury is methylated (or converted into methylmercury) by microbes. In such environments microorganisms take in methylmercury, then small fish eat the microorganisms, and so on. At each step of the food chain methylmercury biomagnifies or accumulates at higher concentrations in an organism. Biomagnification results in situations in which concentrations of a substance found in an organism are higher than could be accounted for by direct exposure alone and the concentration in an organism is higher than is present in the organism's food (USGS, 2010). Large, long-lived fish such as swordfish, shark, and tuna can thus have mercury concentrations in their bodies that are many times higher than those of their surrounding habitat. Humans, a predator at the top of this food chain, can consume toxic amounts of methylmercury when they regularly eat contaminated fish or shellfish. In fact, fish and shellfish consumption is the primary way people in the United States are exposed to methylmercury (Gerstenberger et al., 2010; EPA, 2010). Whether methlymercury in fish is the result of human or natural activities, though, the end result is the same: regular exposure to high doses of methylmercury can harm a human's nervous system.

## Methylmercury, Fish, and Human Health

The health effects of methylmercury exposure vary depending on exposure dose, an individual's rate of absorption and elimination of the compound, and an individual's reaction to the toxicant. The body might eventually convert methylmercury into inorganic mercury and eliminate it in urine and feces. The half-life of mercury in the human body is approximately 70–80 days, "if no additional mercury is ingested" (EPA, 2007, 2–8). While methylmercury is still in the body, it crosses the blood–brain barrier and can affect the nervous system. The severity of the health effects one experiences are related to the level of exposure. Developing nervous systems (such as the ones found in fetuses, infants, and young children) are particularly vulnerable.

High doses of methylmercury exposure are very dangerous. People who consume high doses of methylmercury may have trouble walking or talking and experience delirium, seizures, or death. Children exposed to high levels of methylmercury while *in utero* may be born with severe, irreversible disabilities such as mental retardation, physical deformities, a lack of control over motor skills, or cerebral palsy (Diez, 2009).

Although high-dose poisoning is always a concern, the more common type of exposure in the United States is sustained low-dose exposure. Children who are exposed *in utero* or as young children may have negative changes in cognitive thinking, language, attention, memory, and fine motor and visual spatial skills (NRC, 2000). Such effects can occur whether or not the mother or father displays symptoms of methylmercury poisoning. Symptoms of exposure for the general population include loss of peripheral vision; feelings of numbness and tingling in the face and limbs; loss of coordination in walking; slurred speech; loss of hearing; lowered intelligence; forgetfulness; and muscle weakness (Moore, 2009, 38). Although both high and low doses cause harm, the exact lower dose at which detrimental effects can be seen is not known. Therefore lower-dose limits are open to debate, and different estimates are given as to how much methylmercury should be consumed.

Knowledge about the dangers of methylmercury consumption via seafood may seem commonplace now; however, it took decades of research by numerous teams and individuals across the globe to produce a body of knowledge about methylmercury in fish and its potential effects on humans when consumed.

## Building Awareness of Methylmercury in Fish: A Social Problem in the Making

It took decades and several high-profile events to transform methylmercury hazards into a public issue. During the twentieth century, methylmercury was used as fungicide on seed. It was known to be poisonous; seeds coated in the compound were often an unusual color (for example, pink) to alert consumers. Despite such color coding precautions the color code key was not always known, and at times hunger prevented people from adhering to the danger warnings. Consequently there were high-profile cases of poisoning that received media attention. For example, outbreaks of mercury poisoning occurred in Iraq in 1960, Guatemala in 1963–65, Pakistan in 1969, and Iraq in 1970–71 (Moore, 2009, 41). In the 1970–71 Iraq case some estimate that 5,000 people eventually died from exposure while as many as 50,000 people may have been severely poisoned. In the United States individual cases of methylmercury poisoning made national headlines when a family ate hogs fed with coated grain and the children became severely ill and disabled (Montague and Montague, 1971). Swedish birdwatchers also began documenting high death rates among seed-eating birds such as pheasants, finches, and pigeons and tied this phenomenon to the methylmercury-coated seeds (Montague and Montague, 1971).

In Japan, meanwhile, two high-profile cases catapulted awareness of methylmercury in fish and shellfish into the public eye. Beginning in the 1950s citizens living around Minamata Bay began to report neurological symptoms such as the inability to talk or walk and the experience of delirium. Children born during this time had abnormally high rates of cerebral palsy, mental retardation, and physical deformities. Animals in the area (e.g., cats, fish, birds) also showed signs of trouble. For example, cats walked erratically, had seizures, collapsed, and drowned in the bay. People were initially baffled about the cause of these health problems, but citizens aligned with research scientists to investigate possible causes. At the time the presence of methylmercury in fish and its link to human health were unknown (Saito, 2009). While this effort took place another poisoning outbreak occurred in Niigata Prefecture in 1965.

After years of research and protest the link between consumption of contaminated fish and illness and death was documented, measured, and officially recognized. Two chemical plants—Chisso Corporation Ltd. and Showa Denko Co. Ltd. —were found to be the main sources of mercury pollution (Mishima, 1992). Both plants discharged water containing mercury into the rivers and sea. In 1968 the Japanese government officially recognized that Minamata Disease (the name given to methylmercury poisoning) was caused by the consumption of fish and shellfish contaminated by the chemical plants' discharges (Environmental Health Department, 2002).* Today approximately 3,000 people are officially recognized as victims, while more than 10,000 people have received compensation. Lawsuits against the companies and the government for both the contamination and lack of action continue to be litigated.

During the 1960s and 1970s scientists across the world began testing fish and shellfish to examine the extent of methylmercury contamination. They developed and used a variety of methods to extract tissue and measure methylmercury levels. Sweden, Japan, Canada, and the United States were at the forefront of this effort (Hightower, 2009; Montague and Montague, 1971). Although scientists understood by the 1960s that methylmercury was found in fish it was still unclear how inorganic mercury from chemical plant discharges was transformed into the methylmercury in fish. Swedish scientific teams were convinced that the answer lay in microorganisms in the sediment and began researching this topic. In North America, Dr. J. Woods, University of Illinois, was inspired by Swedish efforts and began to investigate transformation mechanisms (Montague and Montague, 1971, 52). With funding from the National Science Foundation, Dr. Woods and his laboratory team deduced that microorganisms in the mud converted metallic and inorganic mercury into methylmercury. The methylmercury accumulated in algae that were then consumed by small animals and so on.

It took multiple research teams in multiple countries to build a base of knowledge about both the techniques for sampling methylmercury in fish and the biomagnification process involved as the compound moves up the food chain. Although popular ideas of science draw upon a romanticization of the individual or what might be called a hero mythos—i.e., one that highlights a single inventor who has a eureka moment—scientific research is in fact a collaborative effort that includes multiple teams and networks and communication between them. Science is never the effort of one person or one laboratory. Research teams publish papers, and meta-analysis or review studies are done to see where the numerous studies converge or diverge in findings. As sociologists of science have shown scientific research is best understood as a multisited process with multiple ideas moving forward instead of a singular, linear path of development (see, for example, Frickel, 2004; Joyce, 2006; Reardon, 2005). These ideas are shaped by the political, cultural, and economic contexts of a particular society, and in a reciprocal fashion they shape these contexts in turn.

* See Chapter 8 of this volume for a discussion of representations of Minamata in film.

## FDA Enters the Scene

FDA is the US agency charged with stopping the sale of adulterated and misbranded food and drug.* Until the 1970s FDA was primarily a law enforcement agency that employed inspectors who checked for such adulteration and misbranding (Subcommittee on Science and Technology, 2007, 12).

As the methylmercury-in-fish issue got more publicity both within the United States and worldwide, FDA, after internal discussions and data review, issued an interim policy saying that 0.5 parts per million (ppm or μg/g) of mercury in fish was acceptable for the sale of commercial fish. Anything over that should not be sold in US markets (Montague and Montague, 1971). In 1974, FDA's interim policy was converted to an action level, which is a regulatory category that allows the agency to remove fish exceeding the specified limit from grocery stores and other commercial outlets (FDA, 1974). To set an action level harm must be shown. In this case data ranging from case studies of individuals who developed lethargy, headache, difficulty reading and writing, difficulty walking, and hand tremors after regularly eating fish high in methylmercury, to research from Sweden and other countries, results of methylmercury in fish sample tests, and testimony from scientists and physicians were used to formulate the action level (Committee on Commerce, 1971, 93–95). This action level aimed to address the harmful effects that might occur from continued consumption of contaminated fish; its goal was not to protect people from one-time exposures.

Concerns about the levels of methylmercury in tuna (as well as swordfish) were visible in these early discussions. In testimony in front of the US Senate's Committee on Commerce (1971, 105), for instance, the following exchange occurred:

> Mr. Bickwit: Do many weight watchers eat two-and-a-half ounces of fish a day?
>
> Dr. Herdman: I think they do. I would suspect they do. For example, a weight watcher who doesn't like fish, she knows she should eat fish because that is how you lose weight on the program, so she looks around for a fish that doesn't taste too fishy. It happens that she has enough funds to afford a swordfish diet, but if you don't like fish and are looking around for a nonfishy fish, I think tuna fish would be a very logical choice that would combine taste and price for a person who wanted to lose weight.
>
> Mr. Bickwit: A fish that had 0.4 parts per million in it that was consumed by a weight watcher—about how many ounces of this fish would that weight watcher have to eat a day in order to reach dangerous levels?
>
> Dr. Herdnan: To get toxic levels ... should be about 26 ounces which would be a little over a pound and a half in a whole day. You could do that, but for safe levels you could eat about 3 ounces.

As knowledge about the dangers of fish consumption increased, industry pushed back. FDA's action level was increased to 1.0 ppm in 1979 after the agency was successfully sued by Anderson Seafoods Inc., a US swordfish distributor, who argued that the economic impacts of that limit would be detrimental to its business (EPA, 1997b, 6–56; FDA, 1979). The judge sided in favor of Anderson Seafoods,

* See www.fda.gov for the agency's mission statement.

noting the economic harm of the original action level outweighed the harm to individuals' health.

## The Environmental Protection Agency Weighs In

In 1970, President Richard Nixon officially proposed EPA, and it began operation on December 2, 1970, when Congress passed its establishment (Lewis, 1985). As part of its mandate over national waters, EPA received jurisdiction over waterways and recreational and sport fishing in the United States, whereas FDA regulated the commercial fish market.

It was much later, in 1985, though, when EPA produced its first reference dose (RfD) for methylmercury in fish, an RfD of 0.3 μg of methylmercury/kg of one's weight per day (EPA, 1997b, 6–19). An RfD is an estimate of the maximum daily exposure that a person (including one in sensitive subpopulations) can consume without risk of adverse effects, assuming such consumption occurs regularly over a long period (Moore, 2009, 57). This term is likely to be used in risk assessment and toxicology. In contrast, FDA had developed and promoted a tolerable daily intake (TDI), a guideline more likely to be used by human health organizations such as the World Health Organization or Health Canada.* The TDI was similar to an RfD in that it estimated the amount of methylmercury that can be consumed daily over a long period of time with a reasonable certainty of no harm (Carrington and Bolger, 2010). FDA's TDI was 0.3 mg of total mercury/week per person, of which no more than 0.2 mg should be present as methylmercury (EPA, 2000, 5). Thus, EPA recommended 0.3 μg of methylmercury/kg of one's weight per day, whereas FDA recommended 0.47 μg/kg per day. At this moment, EPA and FDA recommendations were not that far apart.

## The Two Agencies Diverge: 1990–2002

During the 1990s, FDA continued to promote the action level for methylmercury it had established decades earlier: 1 ppm.† Despite being publicly reprimanded by the Institute of Medicine, a branch of the National Academies, for its lack of investment in risk assessment practices, FDA did little to test commercial seafood for methylmercury or other toxicants (IOM, 1991; USGAO, 2004). Nor did the agency adequately educate the public or health care professionals about ways to avoid methylmercury when purchasing and eating fish.

For the most part, FDA assumed that methylmercury in fish was not a problem in the United States, in part because officials, drawing on national surveys, believed most people ate very little fish (one serving per month) and tended to choose seafood likely to be low in methylmercury (e.g., shrimp). Lack of fish consumption was actually a bigger concern for the agency, given the potential health benefits associated with eating seafood.

---

* For an overview of definitions of and changes in risk assessment language (e.g., differences between allowable daily intake, tolerable daily intake, and reference dose, safety levels versus uncertainty levels), see Carrington and Bolger, 2010.

† FDA changed its action level for mercury to methylmercury in 1984 (49 Federal Register 45663, 1984).

To address this priority, FDA public affairs staff emphasized the benefits of including fish in one's diet even when writing about the dangers of consuming methylmercury or other contaminants (see, for example, Foulke, 1994; Formanek, 2001).

FDA did, however, publish advisories on eating methylmercury in fish. In 1994 the agency published an advisory in *FDA Consumer* (Foulke, 1994). It advised pregnant women and women of childbearing age who may become pregnant to limit their consumption of shark and swordfish to no more than once a month. People other than pregnant women and women of childbearing age who may become pregnant were advised to limit "consumption of fish species with methylmercury levels around 1 part per million (ppm)—such as shark and swordfish—to about 7 ounces per week (about one serving) to stay below the acceptable daily intake for methylmercury. For fish with levels averaging 0.5 ppm, regular consumption should be limited to about 14 ounces per week" (Foulke, 1994, 8). This notice stated that tuna steaks have tested over 1 ppm, but canned tuna (albacore or light) could be eaten in an unrestricted manner.

In 2001 the agency released a new advisory that simultaneously strengthened and weakened the limits in the warning. On the one hand, the advisory was more protective than the 1994 advisory because it advised pregnant women and women planning on becoming pregnant to avoid shark, swordfish, tile fish, and king mackerel, not consume them once per month (FDA, 2001). On the other hand, it was less protective because people outside of these two groups were not addressed at all. Discussion of canned tuna (or any mention of tuna) did not occur.

During these same years, EPA was diverging in its view of the dangers of fish consumption. Section 112(n)(1)(B) of the Clean Air Act Amendments (CAA), passed in 1990, required EPA to further study the effects of mercury air pollution on wildlife and human life in detail:

> The Administrator shall conduct, and transmit to the Congress not later than 4 years after the date of enactment of the Clean Air Act Amendments of 1990, a study of mercury emissions from electric utility steam generating units, municipal waste combustion units, and other sources, including area sources. Such study shall consider the rate and mass of such emissions, the health and environmental effects of such emissions, technologies which are available to control such emissions, and the costs of such technologies.

Throughout the 1990s, EPA conducted research and meta-analysis of existing data to fulfill the requirements of the Clean Air Act Amendments of 1990. EPA then published, in 1997, an eight-volume report assessing the magnitude of US mercury emissions by source; the health and environmental impacts of those emissions; and the availability and cost of control technologies. This work reviewed existing animal model studies, data from the Japan and Iraq cases, toxicology laboratory studies, and other types of evidence to better understand the potential impacts of mercury on health. In the midst of this review (and as a result of new data), EPA adjusted its reference dose from 0.3 to 0.1 μg/kg of one's weight per day in 1995, thereby becoming more protective about exposure (EPA, 2000, 1). This limit was well below that set by FDA.

One way to test a person's mercury levels is to measure the mercury in their hair or blood. The concentrations of mercury that correspond to EPA's RfD are a mercury blood level of approximately 5 μg/L of blood and a hair mercury concentration of approximately 1.1 μg/kg (EPA, 1997c, 5–26). According to EPA's evaluation humans need to keep their mercury blood and hair concentrations below these amounts in order to avoid adverse health effects. Estimates of hair and blood concentrations in the general US population were scarce in the 1990s, but there were existing blood and hair samples from some subpopulations. Testing of the mercury concentrations in these samples showed that groups linked with high fish consumption frequently exceeded the recommended mercury concentrations (EPA, 1997c). EPA also used published food consumption surveys to estimate that approximately 7% of women nationwide exceeded the RfD (1997c, 6–29). According to EPA, people who exceeded or were estimated to exceed the mercury limit consumed more fish per week than the average US consumer; such individuals were at risk for adverse health effects (e.g., difficulty walking, reading, or talking, dizziness, tremors) associated with methylmercury exposure.

Overall, however, EPA's report affirmed that "the typical US consumer eating fish from restaurants and grocery stores is not in danger of consuming harmful levels of methylmercury from fish and is not advised to limit fish consumption" (1997a, O-2). The assumption about a "typical consumer" (how much fish and what types) came from two national surveys that suggested most Americans eat fish and shellfish (e.g., salmon, shrimp) that are low in mercury, and that they have only one fish meal per month.* In this way, despite setting much lower limits, EPA did concur with FDA.

## Controversy over EPA's Report

EPA's findings caused a stir even before they were published. The eight-volume assessment found itself in an eighteen-month draft limbo, with little political will to make the findings public. Citizens lobbied their federal representatives to get the volumes published while electric utility companies and some fishing corporations quickly challenged EPA's findings (Warrick, 1997). Using the claim industries commonly evoke when their activities are curtailed, they argued that the report was not based on sound science. In response Congress ordered EPA to fund a National Research Council (NRC) panel to determine whether EPA's RfD was based on the best available science (EPA, 2005).

The NRC, which is part of the National Academies, advises the US government on scientific issues. The NRC's Committee on the Toxicological Effects of Methylmercury evaluated new data—epidemiological studies, animal model studies, and the like—that had become available since EPA had done its assessment (NRC, 2000). The committee determined that EPA's reference dose was scientifically sound and should continue to be used as the guideline for protecting public health. They also noted that there were a significant number of women in the United

* EPA used the Continuing Surveys of Food Consumption by Individuals for the periods 1989–91, 1994 and 1995, and the third National Heath and Nutrition Examination Survey (NHANES III) conducted between 1988 and 1994. For more information on the findings from these surveys, see EPA 1997, Executive Summary, 3-37 through 3-39.

States who exceeded the blood and hair levels for methylmercury, corresponding to EPA's RfD. The committee estimated "over 60,000 children are born each year at risk for adverse neurodevelopmental effects due to in utero exposure to methylmercury" (2000, 327).

By the early 2000s, FDA and EPA had clearly diverged. FDA had its action level for methylmercury, which applied to commercially sold seafood. At 1 ppm, it was twice EPA's action level of 0.5 ppm, which applied to recreational fish caught in US local waters. Beyond difference in action levels, FDA invested few resources in studying methylmercury in fish and shellfish and did little to investigate or promote dietary recommendations. On its website during this time, for example, FDA recommended pregnant women and women planning to become pregnant limit their consumption of some fish known to be high in methylmercury, but offered no other easily understood tools for how to think about one's personal exposure to methylmercury in seafood. FDA also maintained the tolerable daily intake of 3.3 μg/kg per week for methylmercury formulated in the 1970s, yet this recommendation was not easy to find among federal documents or protocols. Nor was information about FDA practices and recommendations regarding human health (e.g., mercury blood or hair concentrations) easy to locate.

In contrast, EPA explicitly formulated and reformulated a reference dose for human methylmercury consumption, becoming more protective as more and better data were gathered. EPA further made its reference dose available, publishing it (as well as discussions of how it was developed) in easily accessible documents. EPA offered explanations of human health effects related to mercury exposure, and of the methylmercury cycle, in the mercury section of their website, and clearly stated acceptable mercury blood and hair concentrations.

The press given to EPA's eight-volume report and the NRC report coupled with high-profile clinical studies helped bring the discrepancy between the two agencies to light (see, for example, Hightower and Moore, 2003; Hightower, 2009). State authorities increasingly issued fish advisories warning people to avoid or limit eating contaminated fish from lakes, rivers, and streams (Fairley, 1997). Moreover, some individual states (for example, Hawaii) found that FDA's fish consumption advisory did not adequately protect their population and offered their own, while other states such as Alaska ignored even FDA's recommendation and told everyone to eat as much fish as they wanted (FDA, 2002b).

## Creating a Unified Guideline: FDA and EPA Issue a Joint Fish Advisory

In 2002, FDA's Food Advisory Committee recommended that the two agencies analyze existing scientific research and create a consensus around fish consumption recommendation.* The National Fisheries Institute, National Food Processors Association, the US Tuna Foundation, Science for the Public Interest, the Mercury

* FDA Food Advisory Committee's meeting minutes are transcribed and available to the public on FDA's website. FDA Food Advisory meeting on methylmercury occurred on July 23–25, 2002. The meeting was held at the Sheraton College Park Hotel, Beltsville, Maryland.

Policy Project, the World Health Organization, and other federal and academic programs all provided briefing information documents for the Food Advisory Committee at this meeting. The Food Advisory Committee specifically requested that the agencies develop a guideline for "canned tuna based on a detailed analysis of the contributions canned tuna makes to the overall methylmercury levels in women" (Acheson and Keehner, 2004, 39).

In 2004, EPA and FDA issued the first-ever joint consumer advice about methylmercury in fish and shellfish (Appendix I, Figures 5.1, 5.2). The two agencies collaborated and put forth the advisory, which was published on the FDA and EPA websites and distributed to physicians' offices but not required to be distributed in grocery stores.* The advisory addressed women of childbearing age, pregnant women, and small children; its aim was to keep these individuals' methylmercury hair and blood levels below EPA's reference dose. Yet as FDA Chief Medical Officer David Acheson noted, "The guide is an advisory that we are trying to move people toward. We know that some will be a little over that but we accept that" (2004, 52).

Due in part to an effort to keep the information simple, the advisory did not provide much specific information. The age range for "small children" was not defined. Nor were any of the other groups known to be at risk because of high fish consumption discussed or explicitly addressed (e.g., Native Americans, Pacific Islanders, subsistence fishermen, and women). And nowhere did the document explain that the reference dose was related to one's weight. Instead it made it sound like the advisory applied to all women equally, regardless of size.

The discussion of tuna, a category that includes many species of fish, was particularly confusing. The pamphlet distinguished between three kinds of tuna: canned light tuna, albacore "white" tuna, and tuna steaks. Canned light tuna could be eaten more often than albacore tuna, with recommendations of 12 ounces of canned light tuna per week compared to 6 ounces of albacore tuna per week. In the pamphlet's question-and-answer section, tuna steaks made an appearance: pregnant women or children, it stated, "may eat up to 6 ounces (one average meal) of tuna steak per week." Even though FDA has official definitions of light and white canned tuna, neither category—canned light or albacore white—was defined or explained in the advisory. Nor did it explain why albacore tuna and tuna steaks were less safe than canned light tuna. In fact, the advisory made it seem like there were two types of tuna, albacore tuna and tuna, while in truth there are many species of tuna and some, such as the huge bluefin tuna, had tested higher than both EPA and FDA's action levels for methylmercury. But bluefin tuna did not make it into the high-level-of-contamination, do-not-eat fish category (Ellis, 2008).

Finally, the advisory only mentioned ten fish: shark, swordfish, king mackerel, tilefish, albacore tuna, canned light tuna, shrimp, salmon, pollock, and catfish. It did not list other fish people may encounter in stores or restaurants, nor did it clearly explain

* There are excellent articles that talk about how to communicate the risks associated with methylmercury and how to address diverse publics. See, for example, Burger, 2000, 2005; Burger et al., 2003; Chess et al., 2005; and Groth, 2010.

the expected methylmercury levels of each fish species. It also did not include some fish documented to have high levels of methylmercury (Roe and Hawthorne, 2005).*

The 2004 advisory was a document of compromise. EPA changed its previous emphasis. Subpopulations likely to be at risk from regular fish consumption were not addressed. By focusing on women, pregnancy, and children, the brochure created the idea that mercury levels in fish is an issue solely for women expecting to bear children, pregnant women, and small children, giving the impression that men, teenagers, women not planning on having children, and postmenopausal women need not think about the health risks of eating mercury-laden fish. The distinction between different categories of fish, or different species within a category such as tuna, was also lost. FDA adjusted its previous recommendation by agreeing to note that tuna consumption (albeit in its canned or steak form) should be limited (FDA, 2001). It also agreed to use EPA's reference dose to formulate the advisory.

## THE POLITICS OF TUNA

Shortly before the fish advisory was issued a top agency official acknowledged during a public meeting that FDA's reason for classifying canned light tuna as low in mercury was "to keep market share at a reasonable level" (Hawthorne, 2006). Dr. Vas Aposhian, University of Arizona toxicologist and fish advisory committee member, echoed concerns about market share trumping good science when he noted, "We wanted albacore on the list of fish not to eat. We knew that wouldn't happen because of the pressure from the industry, but we certainly didn't think there should be a recommendation to eat (as much as) 6 ounces of albacore." Dr. Aposhian said that the agencies should have advised women of childbearing age, pregnant women, and young children to completely avoid albacore tuna and eat less light tuna than the guidelines suggest (Kaufman, 2004). To protest the tuna politics of the fish advisory, Dr. Aposhian resigned from the advisory panel.

What is the exact difference between albacore and light tuna? Tuna includes many species of fish. Albacore, bigeye, bluefin, skipjack, and yellowfin are a few of the possible kinds of tuna that may be consumed in American markets. Each species has different growth rates and dietary preferences that affect methylmercury biomagnification consequences. Most of the albacore sold in US grocery stores is the larger, older albacore tuna that is caught using long lines off Asia. These fish are old and large and thus likely to be higher in methylmercury.

Of the many species that comprise tuna, only albacore can be labeled "white tuna" when canned in the United States. The remaining species can be sold as "light tuna."† Together yellowfin tuna and skipjack tuna account for most of the world's

* FDA has a webpage, "Mercury Levels in Commercial Fish and Shellfish," that divides fish into three tables: Table 1: Fish and Shellfish With Highest Levels of Mercury; Table 2: Fish and Shellfish With Lower Levels of Mercury; and Table 3: Mercury Levels of Other Fish and Shellfish. These tables do not provide any information about FDA's action level or tolerable daily intake, making them hard for viewers to interpret.

† FDA regulations define chunk, solid, white, and light tuna. FDA uses a color test, the Munsell value, to categorize tuna as light or white. In this scheme tuna labeled "light" cannot be darker than a Munsell value of 5.3. White tuna cannot be darker than a Munsell value of 6.3. FDA's official definitions of chunk and solid refer to a measurement scale of how fish pieces fit through a mesh screen and the percentage of allowable flake.

canned light tuna, but chunk light may refer to a number of species: bigeye, yellowfin, skipjack, or tongol tuna (Monterey Bay Aquarium, 2003). Thus, light tuna likely will be comprised of yellowfin or skipjack, but it is not guaranteed. As a group tuna comprises more than one-third of the total fish and seafood market and more than half of the finfish market in the United States (National Fisheries Institute, 2010b). Canned tuna is the second most purchased seafood product in the United States after shrimp (FDA, 2002a). High consumption rates combined with the range of methylmercury levels in various species means that tuna (in all its forms) may account for approximately 30–40% of people's mercury consumption from fish and shellfish in the United States (Roe and Hawthorne, 2005b; Groth, 2010; Gerstenberger et al., 2010).

## Testing Tuna

The high rates of tuna consumption in the United States, the lack of attention to the various types of tuna, and the placement of tuna in the low-risk group in the 2004 advisory further mobilized scientists, nonprofit environmental organizations, and journalists. These social actors immediately challenged the 2004 advisory by asking, "Where's the evidence that tuna is safe?" These groups believed the agency's efforts at testing and securing the commercial fish supply were inadequate (see FDA 2004 for a summary of fish tested at this time). Journalists were quick to point out the FDA's data limits. The agency had only tested four walleye and twenty-four shrimp samples since 1978 (Roe and Hawthorne, 2005b). News programs such as *20/20* reported that "three years ago the FDA stopped testing for methylmercury and now requires the fishing industry to police itself" (Hightower, 2009, 40–1). Journalists and nonprofit organizations' staff lost further faith in FDA because the agency seldom enforced its own action level. In truth the agency rarely seized fish that violated FDA's action limits for mercury nor were regulators inspecting seafood in ports, processing plants, or grocery stores for mercury in a systematic, thorough manner (USGAO, 2004).

Given their lack of satisfaction with FDA's data and action, journalists, scientists, and environmental advocates decided to conduct their own studies of methylmercury levels in tuna and other commercially sold fish. In 2005, for example, the *Chicago Tribune* newspaper bought eighteen samples each of eight kinds of fish, including two types of canned tuna, from supermarket chain stores and fish markets in the Chicago area (Roe and Hawthorne, 2005b). A laboratory at Rutgers University analyzed the fish samples. The results: Some of the tuna steak and canned tuna samples exceeded FDA's action level for methylmercury (1 ppm). If a 140-pound woman consumed 6 ounces of tuna per week, as the advisory suggested, she would be over the recommended level of mercury exposure.

Other findings showed that some popular fish such as swordfish had extremely high concentrations of mercury, whereas some fish such as salmon had low concentrations. High concentrations of mercury were also found in orange roughy and walleye—two fish species not mentioned in the 2004 advisory (Roe and Hawthorne, 2005b).

In another study Shawn Gerstenberger and colleagues at the University of Nevada Las Vegas (2010) analyzed three brands of canned tuna—white and light. Chunk white tuna and solid white tuna were both found to have significantly ($p<0.001$) higher concentrations in average methylmercury than chunk light tuna, and one canned tuna brand tested much higher for mercury than the other two. Overall 55% of all tuna tested was above EPA's action level (0.5 ppm) and 5% of the tuna exceeded FDA's action level (1.0 ppm) (Gerstenberger et al., 2010, 238).

Many other studies also tested mercury levels in tuna and other fish and shellfish (see, for example, Burger and Gochfeld, 2006; Burger et al., 2005; Burros, 2008; Lowestein et al., 2010; Sunderland, 2007). Such studies have consistently shown that bluefin varieties of tuna are high in methylmercury and canned albacore tuna, if eaten at FDA and EPA's recommended amounts, could exceed recommended methylmercury levels.

Testing results were published in peer-reviewed journals and newspapers. Reports, with names such as "Toxic Tuna: Mercury Contamination in Chicago Restaurant Tuna Sushi" and "Is Our Tuna Family Safe? Mercury in America's Favorite Fish," were put out by organizations such as Mercury Policy Project, the Got Mercury? project, and similar organizations dedicated to addressing mercury pollution.

## Eat Tuna: Don't Worry

Facing declining sales even before the fish advisory was made public, the tuna industry fought back (NOAA, 2002; Pienin, 2003; Burros, 2006). The US Tuna Foundation and the American Council of Tuna were two organizations that represented tuna companies at the time of the advisory. These groups sought to rebrand tuna as a wonder food, one full of omega-3 fatty acids and other vital nutrients that would produce healthy brains and bodies. They funded other organizations to launch advertising campaigns, develop websites, and provide experts and speakers to news media.

The Tuna Foundation initially gave $45,000 to the University of Maryland's newly formed Center for Food, Nutrition, and Agriculture Policy to create the now defunct website realmercuryfacts.org (Burros, 2006). The Tuna Foundation also donated approximately $500,000 to Harvard Center for Risk Analysis to conduct a benefit-risk analysis of fish consumption (Burros, 2006).

The Center for Consumer Freedom (CCF), a nonprofit organization that primarily receives its funding from fast food, meat, and beverage industries, launched an advertising campaign and websites aimed at discrediting concern about methylmercury and promoting tuna sales (Mayer and Joyce, 2005). CCF sponsors the Fish Scam website that says the risks from mercury are theoretical, and hosts the Mercury Facts website that aims to discredit any claim that suggests methylmercury is risky for humans or that people should consider methylmercury when making fish choices. Listed as a .org site, Mercury Facts encourages viewers to think the site is not tied to industry (.com) interests.

In 2007 the US Tuna Foundation was dissolved and a new council was created within the National Fisheries Institute (NFI) called the Tuna Council. The Tuna Council now lobbies on numerous issues from fishing access arrangements and federal and state regulations to marketing. Beyond direct lobbying efforts via the Tuna

Council, NFI also sponsors two websites: Healthy Tuna (www.healthytuna.com) and About Seafood (www.aboutseafood.com). Both sites promote the idea that tuna is a healthy, safe source of nutrients and downplay the risks associated with frequent tuna consumption. To further brand tuna as a food of the elite, About Seafood offers recipes for meals such as grilled tuna with fruit salsa and pepper seared yellowfin.

The Edison Electric Institute, meanwhile, an association of shareholding electric companies, supports a website, www.mercuryanswers.org, that promotes the idea that the electric companies are doing the most they can about mercury pollution and that most people can consume fish without concern about methylmercury. Mercury Answers proposes:

> For most people, the risk from mercury by eating fish and shellfish is not a health concern, according to the U.S. Environmental Protection Agency and the Food and Drug Administration. However, just to be safe, these government agencies advise women who may become pregnant, pregnant women, nursing mothers, and young children to avoid some types of fish and to eat fish and shellfish that are lower in mercury.

This strategy— create an organization or website with a pro-environment name; use media campaigns to positively brand your product; select isolated quotes from individuals or organizations instead of looking at the body of scientific data; make it seem like there is debate when there is consensus among scientists—is one often used in health or environmental disputes. In all, these websites send a clear message: eat lots of tuna and other seafood, and don't worry about methylmercury, PCBs, or other contaminants.

The pro-tuna websites also alter the story of how methylmercury enters fish. Instead of saying it is the result of both human and natural processes, such sites emphasize that the presence of methylmercury is primarily natural. The CannedAlbacoreTuna.com website, for example, explains that "mercury is a natural element that is found in minute quantities in air, water and all living things. Some fish have more mercury than others depending on their environment and their position in the food chain." CCF's Mercury Facts website asks, "Did you know? Methylmercury in ocean caught fish is almost entirely naturally occurring." Although natural sources are thought to be the main sources of methylmercury in ocean fish (Kraepiel et al., 2003; Sunderland et al., 2009), human sources of mercury pollution are downplayed or ignored outright on these sites. Moreover, "natural" is used as a synonym for "harmless." By creating the idea that methymercury in fish is a natural phenomenon such sites seek to reassure consumers that it is OK to consume methylmercury. They also suggest that consumers or organizations do not have to work to reduce human sources of mercury pollution. Why? Because mercury is natural.

The list of strategies employed by these industry-funded groups goes on and on. CCF created its own methylmercury calculator that uses a more lenient lower limit, EPA's Base Dose Lower Limit (BDLL), than what EPA normally recommends (CCF, 2010). Editorials are written for newspapers and letters to the editor help get their message out (see, for example, Martosko, 2008, 2006). For these actors, fish is a wonder food. The harms associated with consuming it have been overblown.

## One Size Does Not Fit All

While the public battle between the tuna industry, environmental organizations, and federal agencies took place, states and cities took action. Hawaii, for example, provides its residents with more protective advice than the federal government by recommending that adults limit their consumption of canned tuna to 6 ounces (one can) per week and children limit their consumption of canned tuna to 3 ounces (half a can) per week (Hawaii State Department of Health, 2003). The City of Vancouver, Washington, has developed an interactive website to help children learn how to choose fish more likely to have low levels of methylmercury. Although the colorful website is careful to say that knowing how much mercury any one fish has is impossible, it makes clear that tuna is one of five fish to be avoided.* However, the website does not distinguish between types of tuna (e.g., bluefin, albacore), lumping all tuna species into one category.

In California, consumers may find warnings on the risks associated with eating contaminated fish (including tuna) posted in chain restaurants and grocery stores. This is due in large part to Proposition 65, a law drafted by the late Al Meyerhof when he worked for the Natural Resource Defense Council. Proposition 65 requires consumer warnings for disclosure of toxic substances in products. In 2003, drawing on Proposition 65, the California Attorney General's office filed suit to force Safeway, Trader Joe's, and three other grocery chains to warn customers that tuna (fresh, frozen, and canned), swordfish, and shark sold in their markets contain methylmercury (Egelko, 2003). In the same year, the California Attorney General also filed suit against chain restaurants, requiring them to post warnings about methylmercury in fish. In January 2005, most major restaurant chains sued by the state agreed to put up warnings (NOW, 2005). Environmental organizations such as the Turtle Island Restoration Network and Oceana, meanwhile, work with supermarkets to get them to post warnings in seafood sections.

Other nonprofit organizations challenged the lack of specificity about types of fish in the 2004 EPA and FDA fish advisory as well as its limited audience: pregnant women, women who may become pregnant, and small children. Under the banner of "one size does NOT fit all," organizations created their own fish advisories and methylmercury seafood calculators to help individuals assess the risks associated with their seafood consumption. For example, the Environmental Defense Fund publishes its own fish advisory (Appendix II, Figure 5.3). Organizations such as the Sierra Club, the Natural Resource Defense Council, and Physicians for Social Responsibility also have developed printed advisories that list fish by estimated methylmercury levels so that consumers can make educated choices for themselves. The Turtle Island

* The City of Vancouver's interactive site is called "Play Our Tuna Game." It is available at http://www.ci.vancouver.wa.us/solidwaste/pbt_site/images/PBT_tuna_game.swf

Restoration Network, originator of the Got Mercury? campaign, and the Natural Resource Defense Council each respectively developed an online mercury in seafood calculator that uses EPA data to calculate one's potential exposure.* These groups recognize that there are subpopulations of US residents who regularly and frequently eat fish; their aim is to help such individuals understand the link between frequent fish consumption, fish choice, and negative health effects from methylmercury ingestion. Unlike FDA and EPA 2004 food advisory (which does not address these individuals), the environmental nonprofit organizations specifically take them into account.

### Getting the Word Out?

FDA initially distributed more than 4 million brochures (available in Spanish and English) to medical offices, and contacted approximately 9,000 electronic and print outlets and editors of pregnancy books, who were urged to include information about the advisory in upcoming editions (Acheson, 2005). More than fifty organizations of health care providers to women and their families were contacted (Acheson, 2005). FDA also continues to make the advisory available on its website in both Spanish and English.

EPA also made outreach to various publics a priority. In addition to including links to reports and other documents related to mercury in fish, for example, EPA created an interactive website, EPA's Fish Kids, that helps kids (and their parents) learn how to pick fish that are low in methylmercury. The take-home message of this site is: eat more fish, but choose fish that are low in methylmercury.

EPA makes the 2004 EPA and FDA Fish Advisory available in eight languages (online and as hard copies) and developed a "One Fish, Two Fish, Don't Fish, Do Fish" poster, a keychain containing the recommendation to "Eat two meals per week of a variety of fish lower in mercury, such as: canned light tuna, catfish, pollack, salmon and shrimp," and a "Fish for your Health" magnet. These items are available for free and all emphasize that people should eat fish that are low in mercury.

However, federal requirements such as posting information in supermarket aisles or on the labels of canned or packaged fish have certainly never gotten off the ground.

## SCIENCE AS CULTURE, SCIENCE AS POLITICS

Why did FDA and EPA understand the risks associated with methylmercury in fish so differently? Many factors contribute to their divergence in understanding and communication about risks associated with methylmercury consumption. The divergence is not necessarily unusual. Each federal science agency—National Institutes of Health, National Science Foundation, FDA, EPA, Office of Naval Research—has its own institutional structure and culture, way of implementing peer review, and

* The Got Mercury? calculator is available at http://www.gotmercury.org/article.php?list=type&type=75. The Natural Resource Defense Council mercury calculator can found at http://www.nrdc.org/health/effects/mercury/calculator/start.asp. Both calculators use EPA data to formulate methylmercury levels.

type of scientific expertise. Here I highlight four factors that contribute to FDA and EPA's differing understandings of risk.

First, a part of FDA focuses on nutrition and diet for the entire nation. The focus on nutrition and health is central to one of the agency's eight centers—the Center for Food Safety and Applied Nutrition, or CFSAN. Michael Bolger, PhD in physiology and biophysics, and David A. Acheson, MD, both central figures in the FDA's methylmercury assessment, were employed at the center, and it was CFSAN that issued the fish advisories. One of the FDA's mandates is to advance "the public health by helping to speed innovations that make medicines and foods more effective, safer, and more affordable; and helping the public get the accurate, science-based information they need to use medicines and foods to improve their health."* Since fish, minus the methylmercury, is certainly a healthy food to include in a typical American's diet, FDA emphasizes the health benefits of fish at the expense of discussing environmental exposure to methylmercury. Given that most Americans do not eat enough fish, this majority population is privileged in FDA publications and guidelines.

Second, FDA does not have an institutional structure that empowers scientists trained in risk assessment, environmental exposure, or epidemiology, nor does it have a robust peer review process. This point was made clear in a federal report called *FDA Science and Mission at Risk*. Among its major findings it was noted that "FDA cannot fulfill its mission because its scientific base has eroded and its scientific organizational structural is weak" (2007, 3). The subcommittee further highlighted how "FDA lacks sufficient expertise in quantitative methods, such as statistics and biomathematics," and noted that "FDA also has a lack of expertise in risk/benefit assessment" (2007, 31–32). Such forms of expertise (combined with institutional recognition of their importance) are crucial to the evaluation of methylmercury in fish.

Part of the problem stems from the FDA's legacy as a regulatory institution. Until the 1970s, FDA was primarily a law enforcement agency (Subcommittee on Science and Technology, 2007, 12). As such it relied less on scientific research and knowledge, and invested in inspectors who could check for adulteration and misbranding (Hilts, 2004). Although FDA now employs toxicologists and epidemiologists, the agency's current configuration still does not have the scientific staff or peer review practice needed to carefully evaluate risk or environmental exposures. A robust peer review practice invites reviewers of appropriate forms of expertise, has the networks to make the invitations, and incorporates feedback from peer review.

Third, FDA not only reflects a regulatory past but also embraces forms of medical expertise that privilege individualized clinical care and genetics over other kinds of scientific knowledge. Medical education stresses knowledge of genetics, pharmacological treatments, and surgical interventions whereas discussion of the links between toxicants and illness are seldom included (Brown and Mikkelsen, 1990). Environmental exposures are marginal in both medical education and medical practice. This marginality in the broader medical milieu makes it harder for FDA to put environmental concerns front and center.

---

* See http://www.fda.gov/AboutFDA/WhatWeDo/default.htm.

Finally, FDA staff and committee members have a long history of a revolving door policy with the very industries they are supposed to regulate (Cauchon, 2000). From the recalls associated with blockbuster drugs such as Vioxx to the failure to regulate bovine growth hormone, FDA advisers have been charged with being too close to industry (Richwine, 2004). In the case of seafood advisories, the ties between FDA advisors and food industries are evident. For example, Louis Sullivan, former secretary of the US Department of Health and Human Services, supported the move to keep tuna on the "do eat" list and worked to discredit the science showing high levels of methylmercury in tuna. After leaving the US Department of Health and Human services, Sullivan became a consultant for the Tuna Foundation (Hawthorne, 2006). Further illustrating the close ties between the tuna industries and FDA, FDA Commissioner Lester M. Crawford challenged California's lawsuit against Tri-Union Seafoods LLC, Del Monte Corporation, and Bumble Bee Seafoods LLC (California Court of Appeal, 2009). The lawsuit would have required canned tuna to have mercury labels. In support of the tuna companies' antilabeling position, Crawford said that such labels would frustrate the FDA's approach to advise "consumers of both the benefits and possible risks of eating fish and shellfish" (California Court of Appeal, 2009, 9).

In contrast, EPA directly employs scientists trained in risk assessment, toxicology, epidemiology, and other related sciences. These forms of expertise have been the mainstay of the agency whereas medical forms of expertise (medical doctors, dieticians) have been more marginal. The agency relies on academic and industry data, and conducts its own research. EPA also incorporates results from peer review processes; it does not ignore peer review findings or suggestions. In the case of mercury the agency did not have the same level of movement between employees and industries. Key figures such as Kathryn R. Mahaffey, PhD, were career scientists who worked at EPA and then moved to academic appointments if they took any other position at all. The leap between working on protecting the environment and working for a company that pollutes the environment (e.g., utility companies, chemical companies) is perhaps harder to accept in comparison to rotating between the FDA and fish or other food-related industries.

In all, these four factors make it harder for FDA to imagine methylmercury in fish as harmful whereas EPA standards are more protective of human health.

## CONCLUSION: SHOULD WE EAT TUNA?

Tuna is a highly consumed fish in the United States. Although many varieties comprise the label "tuna," this distinction is at times lost in public discussion and in some fish advisories. From sushi to canned, the species of tuna being sold varies. The species matters, however, in terms of methylmercury levels. Some tuna species such as bluefin are more likely to have higher levels of methylmercury than others (e.g., skipjack). The age of the fish when caught also matters. Meanwhile, some studies show that even types of tuna that should be lower in methylmercury are not (see, for example, Groth, 2010; Malsch and Muffett, 2006; Savitz et al., 2008). Lack of regulation regarding imports and minimal testing of the commercial fish supply make it hard to know how much methylmercury is in the tuna one purchases.

FDA and EPA diverge in how they understand and communicate the benefits and risks associated with human consumption of methylmercury in fish. This divergence is related in part to how each agency defines scientific expertise, their implementation of the peer review process, as well as the individuals who have worked on mercury at each agency. EPA, with its emphasis on risk assessment, toxicology, and epidemiology, formulates an action level that is more protective than the FDA's. EPA also aims to communicate information to different publics and makes its reference dose and the fish advisory available to multiple subgroups within the United States. FDA did not choose peer reviewers with expertise in epidemiology, toxicology, or risk assessment, nor did it incorporate criticisms from peer reviewers. Moreover, scientists employed at EPA did not have the same ties to the fish industry as did FDA scientists. These differences help explain why each agency produces varied views on fish consumption.

FDA and EPA came together for the 2004 fish advisory. This alliance, however, is fragile. In 2009, FDA filed drafts of two risk and benefit assessment reports—*Report of Quantitative Risk and Benefit Assessment of Commercial Fish Consumption, Focusing on Fetal Neurodevelopmental Effects (Measured by Verbal Development in Children) and on Coronary Heart Disease and Stroke in the General Population* and *Summary of Published Research on the Beneficial Effects of Fish Consumption and Omega-3 Fatty Acids for Certain Neurodevelopmental and Cardiovascular Endpoints*—that may undermine the weak coalition advisory that currently exists. The drafts' take-home message is that the benefits of eating fish outweigh the risks associated with methylmercury consumption. Instead of formulating a clear "eat more fish, but choose fish that are low in mercury" message, these reports primarily emphasize the benefits of eating more fish.

EPA (2009) offered lengthy comments to FDA during an interagency review, prior to the reports' release for public comment. Although FDA's draft reflects some changes in response to these comments, EPA did not receive formal response from FDA about them (EPA, 2009, 1). EPA subsequently published a lengthy response (sixty-one detailed pages) in FDA's docket dedicated to the draft. The agency's response calls into question everything from FDA's unit of analysis to the agency's discussion of uncertainty. The differences in expertise are apparent in the exchange about the draft's scientific assumptions and methodologies as well as in FDA's choice of peer reviewers and the agency's limited response to peer review criticisms.

Environmental organizations and industry-affiliated organizations also offer fish advice. For environmental organizations, FDA and EPA advisory is not protective enough. It does not address individuals who regularly eat fish, nor does it adequately account for the levels of methylmercury in highly consumed fish. Such groups challenge FDA's emphasis on risk communication to medical professionals and insist that consumers have advice at the point of sale (e.g., through posters in seafood sections, labels on canned tuna) or at the very least in their pockets (e.g., fish advisory wallet-sized cards) when they are shopping. For organizations representing the tuna or electric industries, the 2004 FDA and EPA advisory is too protective. These organizations stress the benefits of eating fish and use examples from science to support their position. Such groups downplay or, in the case of CCF's strategies, ridicule concerns about methylmercury or other toxicants in fish. These strategies in turn obscure the otherwise simple health message: eat fish, but choose fish that is low in methylmercury.

Advisories are social artifacts—ones that reflect forms of scientific expertise, economic priorities, and ideas about how and to whom risk should be communicated. Both high and low doses of methylmercury cause harm, but lower-dose limits are open to debate, and different estimates are given as to how much methylmercury can safely be consumed. Nations, states, and government agencies issue fish consumption guidelines, but how little methylmercury may cause neurological harm is uncertain. Using a sociological lens illustrates the values and the types of science in use in the creation of fish advisories.

# APPENDIX I

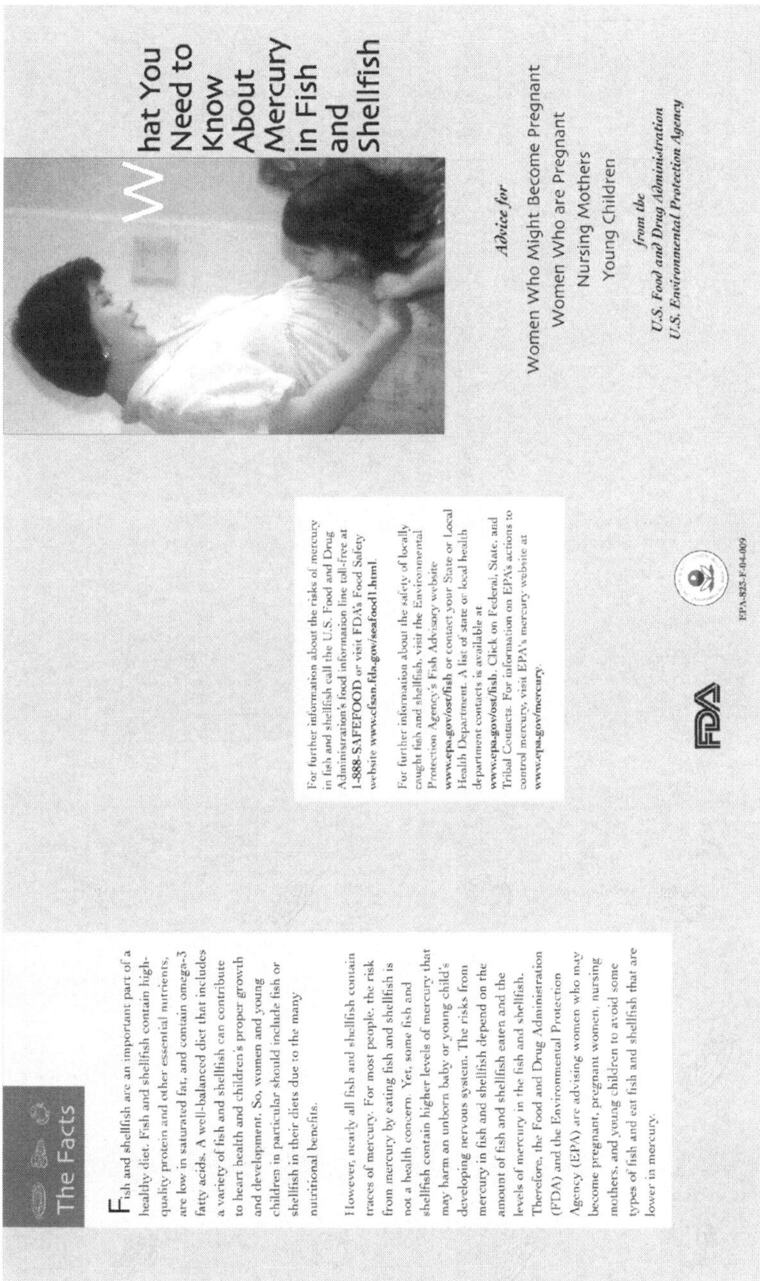

The Facts

Fish and shellfish are an important part of a healthy diet. Fish and shellfish contain high-quality protein and other essential nutrients, are low in saturated fat, and contain omega-3 fatty acids. A well-balanced diet that includes a variety of fish and shellfish can contribute to heart health and children's proper growth and development. So, women and young children in particular should include fish or shellfish in their diets due to the many nutritional benefits.

However, nearly all fish and shellfish contain traces of mercury. For most people, the risk from mercury by eating fish and shellfish is not a health concern. Yet, some fish and shellfish contain higher levels of mercury that may harm an unborn baby or young child's developing nervous system. The risks from mercury in fish and shellfish depend on the amount of fish and shellfish eaten and the levels of mercury in the fish and shellfish. Therefore, the Food and Drug Administration (FDA) and the Environmental Protection Agency (EPA) are advising women who may become pregnant, pregnant women, nursing mothers, and young children to avoid some types of fish and eat fish and shellfish that are lower in mercury.

For further information about the risks of mercury in fish and shellfish call the U.S. Food and Drug Administration's food information line toll-free at **1-888-SAFEFOOD** or visit FDA's Food Safety website **www.cfsan.fda.gov/seafood1.html**.

For further information about the safety of locally caught fish and shellfish, visit the Environmental Protection Agency's Fish Advisory website **www.epa.gov/ost/fish** or contact your State or Local Health Department. A list of state or local health department contacts is available at **www.epa.gov/ost/fish**. Click on Federal, State, and Tribal Contacts. For information on EPA's actions to control mercury, visit EPA's mercury website at **www.epa.gov/mercury**.

FDA

EPA-823-F-04-009

What You Need to Know About Mercury in Fish and Shellfish

*Advice for*

Women Who Might Become Pregnant

Women Who are Pregnant

Nursing Mothers

Young Children

*from the*

*U.S. Food and Drug Administration*

*U.S. Environmental Protection Agency*

**FIGURE 5.1** Joint FDA and EPA fish consumption advisory, page 1.

## 3 Safety Tips

By following these 3 recommendations for selecting and eating fish or shellfish, women and young children will receive the benefits of eating fish and shellfish and be confident that they have reduced their exposure to the harmful effects of mercury.

1. **Do not eat:**
   - **Shark**
   - **Swordfish**
   - **King Mackerel**
   - **Tilefish**

   They contain high levels of mercury.

2. **Eat up to 12 ounces (2 average meals) a week of a variety of fish and shellfish that are lower in mercury.**
   - Five of the most commonly eaten fish that are low in mercury are shrimp, canned light tuna, salmon, pollock, and catfish.
   - Another commonly eaten fish, albacore ("white") tuna has more mercury than canned light tuna. So, when choosing your two meals of fish and shellfish, you may eat up to 6 ounces (one average meal) of albacore tuna per week.

3. **Check local advisories about the safety of fish caught by family and friends in your local lakes, rivers, and coastal areas.**

   If no advice is available, eat up to 6 ounces (one average meal) per week of fish you catch from local waters, but don't consume any other fish during that week.

**Follow these same recommendations when feeding fish and shellfish to your young child, but serve smaller portions.**

Visit the Food and Drug Administration's Food Safety Website **www.cfsan.fda.gov** or the Environmental Protection Agency's Fish Advisory Website **www.epa.gov/ost/fish** for a listing of mercury levels in fish.

## Frequently Asked Questions *about Mercury in Fish and Shellfish:*

### *What is mercury?*

Mercury occurs naturally in the environment and can also be released into the air through industrial pollution. Mercury falls from the air and can accumulate in streams and oceans and is turned into methylmercury in the water. It is this type of mercury that can be harmful to your unborn baby and young child. Fish absorb the methylmercury as they feed in these waters and so it builds up in them. It builds up more in some types of fish and shellfish than others, depending on what the fish eat, which is why the levels vary.

### *I'm a woman who could have children but I'm not pregnant - so why should I be concerned about methylmercury?*

If you regularly eat types of fish that are high in methylmercury, it can accumulate in your blood stream over time. Methylmercury is removed from the body naturally, but it may take over a year for the levels to drop significantly. Thus, it may be present in a woman even before she becomes pregnant. This is the reason why women who are trying to become pregnant should also avoid eating certain types of fish.

### *Is there methylmercury in all fish and shellfish?*

Nearly all fish and shellfish contain traces of methylmercury. However, larger fish that have lived longer have the highest levels of methylmercury because they've had more time to accumulate it. These large fish (swordfish, shark, king mackerel and tilefish) pose the greatest risk. Other types of fish and shellfish may be eaten in the amounts recommended by FDA and EPA.

Note:
*If you have questions or think you've been exposed to large amounts of methylmercury, see your doctor or health care provider immediately.*

### *I don't see the fish I eat in the advisory. What should I do?*

If you want more information about the levels in the various types of fish you eat, see the FDA food safety website www.cfsan.fda.gov/~frf/sea-mehg.html or the EPA website at www.epa.gov/ost/fish.

### *What about fish sticks and fast food sandwiches?*

Fish sticks and "fast-food" sandwiches are commonly made from fish that are low in mercury.

### *The advice about canned tuna is in the advisory, but what's the advice about tuna steaks?*

Because tuna steak generally contains higher levels of mercury than canned light tuna, when choosing your two meals of fish and shellfish, you may eat up to 6 ounces (one average meal) of tuna steak per week.

### *What if I eat more than the recommended amount of fish and shellfish in a week?*

One week's consumption of fish does not change the level of methylmercury in the body much at all. If you eat a lot of fish one week, you can cut back for the next week or two. Just make sure you average the recommended amount per week.

### *Where do I get information about the safety of fish caught recreationally by family or friends?*

Before you go fishing, check your Fishing Regulations Booklet for information about recreationally caught fish. You can also contact your local health department for information about local advisories. You need to check local advisories because some kinds of fish and shellfish caught in your local waters may have higher or much lower than average levels of mercury. This depends on the levels of mercury in the water in which the fish are caught. Those fish with much lower levels may be eaten more frequently and in larger amounts.

**FIGURE 5.2** Joint FDA and EPA fish consumption advisory, page 2

## APPENDIX II

A ▼

SEAFOOD
SELECTOR
Fish choices that are good for you and the ocean

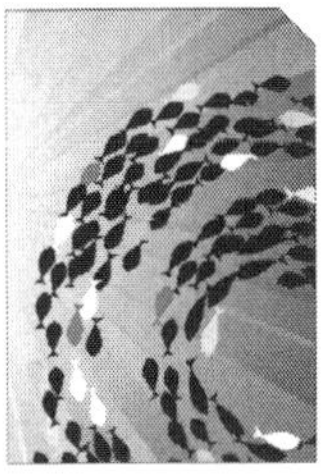

EDF ENVIRONMENTAL DEFENSE FUND

For more information, visit:
EDF.org/seafood

From your mobile phone:
m.EDF.org/seafood

EDF ENVIRONMENTAL DEFENSE FUND

This guide is produced in collaboration with the Monterey Bay Aquarium (seafoodwatch.org).

©February 2010 Environmental Defense Fund. Printed on recycled paper (30% post-consumer), totally chlorine free.

Finding eco-friendly fish
When you're at the fish counter or restaurant, knowing what questions to ask is key. Your server or fishmonger may not have all the answers, but the more you ask, the more they will recognize the need to provide better information to their customers. Here are some easy questions to get you started:

- Is this fish wild-caught or farm-raised?
- What country or geographic area is it from?
- If it is farmed, how was it grown?
- If it is wild, how was it caught?

This will not only help you follow the recommendations on the other side of this card; it also sends a powerful conservation message that you'd like to see more sustainable seafood choices in the case or on the menu. And finally, don't be afraid to spread the word to friends and family. The more people who ask for and buy ocean-friendly, the better!

Your seafood choices matter!
Not all seafood is created equal. Unfortunately many popular seafood items are fished or farmed in environmentally destructive ways. Because these practices vary by species and geographic location, knowing this information makes a big difference and can guide your decisions when it comes to ordering fish. By using your purchasing power to buy fish caught or farmed using eco-friendly practices, you are supporting healthy abundant oceans.

How does this card work?
Seafood can be an Eco-Best, Eco-OK or Eco-Worst depending on where it came from and how it was caught or farmed. Our guide helps you make the best environmental choice for the oceans. We've also used green hearts and red triangles to indicate the fish that offer the greatest health benefits (e.g. high in omega-3s) or risks (e.g. high in mercury or PCBs), respectively.

C ▶ ◀ C

B ▶ ◀ B

| FISH | ECO-BEST | ECO-OK | ECO-WORST |
|---|---|---|---|
| Arctic char | Farmed ♥ | | |
| Catfish | U.S. | Basa/swai/tra | |
| Caviar/sturgeon | | U.S. Farmed | Imported or wild |
| Chilean seabass | | | All▲ |
| Clams | Farmed | Wild | |
| Cod | Pacific (Alaska longline) | Pacific (U.S. trawl) | Atlantic, Pacific (imported) |
| Crab | Dungeness, Stone | King (U.S.), Snow, Blue▲ | King (imported) |
| Crawfish | U.S. | | Imported |
| Flounder/Sole | | Pacific | Atlantic |
| Grouper | | | All▲ |
| Haddock | | Hook and line | Trawl |
| Halibut | Pacific | | Atlantic |
| Herring | | Atlantic | |
| Lobster | Spiny (U.S.) | American/Maine, Spiny (Bahamas) | Spiny |
| Mahimahi | U.S. (Atlantic pole/troll) | U.S. (other) | Imported |
| Monkfish | | | All |
| Mussels | Farmed | | |
| Orange roughy | | | All▲ |
| Oysters | Farmed ♥ | Wild ♥ | |
| Pollock | | All | |
| Rockfish | | Pacific (hook and line)▲ | Pacific (trawl)▲ |
| Sablefish/Black cod | Alaska, Canada ♥ | California, Oregon, Washington ♥ | |
| Salmon | Alaska (wild) ♥, Canned (sockeye, pink) ♥ | Washington (wild) ▲ | Atlantic <u>or</u> farmed▲ |
| Sardines | U.S. ♥ | | |
| Scallops | Bay (farmed) | Sea (wild) | |
| Shark | | | All▲ |
| Shrimp | Pink (Oregon) ♥, Spot prawns (Canada) ♥ | U.S. or Northern ♥ | Imported |
| Skate | | | All |
| Snapper | | Gray, Lane, Mutton (U.S.) | Red, Imported, Vermillion |
| Squid | Longfin (U.S.) | All other | |
| Striped Bass | Farmed <u>or</u> wild | | |
| Swordfish | | U.S.▲ | Imported▲ |
| Tilapia | U.S. | Latin America | Asia |
| Trout | Rainbow (farmed) | | |
| Tuna | Albacore (Canada, U.S.) ♥, Bigeye, or Yellowfin (U.S. pole/troll), Skipjack (pole/troll) | Canned (light/skipjack or white/Albacore)▲, Bigeye or Yellowfin (imported pole/troll or U.S. longline) | Albacore, Bigeye, or Yellowfin (imported longline)▲, Bluefin▲ |

♥ A good source of heart-healthy omega-3s <u>and</u> low in contaminants
▲ Limit consumption due to elevated mercury or PCB levels

More details at EDF.org/seafood

▲ A

1. Cut along dotted line.
2. Fold in half along A, keeping printed side of paper on outside.
3. Fold in half along B, keeping fish list on inside.
4. Fold in half along C, keeping Seafood Selector panel in front.

**FIGURE 5.3** Environmental Defense fund fish advisory.

## REFERENCES

Acheson, David. 2005. Implementation of the FDA/US EPA joint advisory. Pp. II-77-II-78. In *Proceedings of the 2004 National Forum on Contaminants in Fish*. Baltimore, MD: September 18–25. EPA 823-R-05-006.

Acheson, David, and Denise Keehner. 2004. National mercury advisory: Description of existing advisory and August 2003 FDA FAC recommendations. In *Proceedings of the 2004 National Forum on Contaminants in Fish*, 39–41. San Diego, CA: January 25–28. EPA 823-R-04-006.

Brown, Phil, and Edwin Mikkelsen. 1990. *No Safe Place: Toxic Waste, Leukemia, and Community Action*. Berkeley, CA: University of California Press.

Burger, Joanna. 2000. Consumption advisories and compliance: The fishing public and the deamplification of risk. *Environmental Planning and Management,* 43:471– 488.

Burger, Joanna. 2005. Fishing, fish consumption, and knowledge about advisories in college students and others in central New Jersey. *Environmental Research,* 98:268–275.

Burger, Joanna, Melanie Hughes McDermott, Caron Chess, Eleanor Bochenek, Marla Perez-Lugo, and Kerry Kirk Phlugh. 2003. Evaluating risk communication about fish consumption advisories: Efficacy of a brochure versus a classroom lesson in Spanish and English. *Risk Analysis,* 23(4):791–803.

Burger, Joanna, Alan Stern, and Michael Gochfeld. 2005. Mercury in commercial fish: Optimizing individual choices to reduce risk. *Environmental Health Perspectives,* 113(3):266–271.

Burger, Joanna, and Michael Gochfeld. 2006. Mercury in fish available in supermarkets in Illinois: Are there regional differences? *Science of the Total Environment,* 367:1010–1016.

Burros, Marian. 2006. Advisories on fish and the pitfalls of good intent. *The New York Times,* February 15.

Burros, Marian. 2008. High mercury levels are found in tuna sushi. *The New York Times,* January 23.

California Court of Appeal. 2009. *The people ex re. Edmund G. Brown, Jr., as Attorney General, etc., et al., plaintiffs and appellants, v. Tri-Union Seafoods, LLC, et al., defendants and respondents*.

Carrington, Clark, and P. Michael Bolger. 2010. The limits of regulatory toxicology. *Toxicology and Applied Pharmacology,* 243(2):191–197.

Cauchon, Dennis. 2000. FDA advisers tied to industry. *USA Today,* September 25.

Center for Consumer Freedom (CCF). 2010. Seafood Calculator. Accessed May 28, 2010, http://howmuchfish.com/.

Chess, Caron, Joanna Burger, and Melanie McDermott. 2005. Speaking like a state: Environmental justice and fish consumption advisories. *Society and Natural Resources,* 18(3):267–278.

Committee on Commerce. 1971. *Fish Inspection Legislation: Hearings Before the Subcommittee on the Environment of the Committee on Commerce, US Senate. Serial No. 92–16.* Washington, DC: US Government Printing Office.

Diez, Sergi. 2009. Human health effects of methlymercury exposure. In *Reviews of Environmental Contamination and Toxicology*, vol. 198, David Whitacre, ed., 111–129. New York: Springer New York.

Egelko, Bob. 2003. California sues Safeway, Kroger Co., Albertson's, Trader Joe's and Whole Foods over mercury in fish. *San Francisco Chronicle,* January 18.

Ellis, Richard. 2008. *Tuna: Love, Death, and Mercury*. New York: Vintage Books.

Environmental Health Department. 2002. *Minamata Disease: The History and Measures*. Ministry of the Environment, Government of Japan. Godochosha No. 5, 1-2-2 Kasumigaseki, Chiyoda-ku, Tokyo 100–8975, Japan.

Environmental Protection Agency. 1997a. *Mercury Study Report to Congress. Vol. I.: Executive Summary.* EPA-452/R-97-003. US Environmental Protection Agency, Office of Air Quality Planning and Standards, and Office of Research and Development.

Environmental Protection Agency. 1997b. *Mercury Study for Congress. Volume VI: Characterization of Human Health and Wildlife Risks from Anthropogenic Mercury Emissions in the United States.* EPA-452/R-97-008b. US Environmental Protection Agency, Office of Air Quality Planning and Standards, and Office of Research and Development.

Environmental Protection Agency. 1997c. *Mercury Study for Congress. Volume VII: Characterization of Human Health and Wildlife Risks from Mercury Exposure in the United States.* EPA-452/R-97-009. US Environmental Protection Agency, Office of Air Quality Planning and Standards, and Office of Research and Development.

Environmental Protection Agency. 2000. *Reference Dose for Mercury.* External review draft. US Environmental Protection Agency, National Center for Environmental Assessment NCEA-S-0930.

Environmental Protection Agency. 2005. *Derivation of an RfD for Methylmercury.* National Center for Environmental Assessment. Accessed April 31, 2010, http://cfpub.epa.gov/ncea/cfm/recordisplay.cfm?deid=22491.

Environmental Protection Agency. 2009a. EPA Fish Kids website. http://www.epa.gov/fishadvisories/kids/.

Environmental Protection Agency. 2009b. Comments on the FDA draft report, *Report of Quantitative Risk and Benefit Assessment of Commercial Fish Consumption, Focusing on Fetal Neurodevelopmental Effects (Measured by Verbal Development in Children).* Docket No. FDA-2009-N-0018-0138.

Environmental Protection Agency. 2010. *Mercury: Basic Information.* Accessed April 17, 2010, http://www.epa.gov/mercury/about.htm.

Fairley, Peter. 1997. Fish advisories up: EPA cites better reporting. *Chemical Week*, July 30, 14.

Food and Drug Administration. 1974. *Federal Register 39*; 42738, December 6.

Food and Drug Administration. 1979. *Federal Register 44*; 3990, January 19.

Food and Drug Administration. 2002a. Minutes of FDA's Food Advisory Committee meeting on methylmercury. Sheraton College Park Hotel, Beltsville, Maryland, July 23–25.

Food and Drug Administration. 2002b. Transcript of FDA's Food Advisory Committee meeting on methylmercury. Sheraton College Park Hotel, Beltsville, Maryland, July 23–25.

Food and Drug Administration. 2004. Mercury concentrations in fish: FDA monitoring program (1990–2004). Accessed April 25, 2010, http://www.fda.gov/Food/FoodSafety/Product-SpecificInformation/Seafood/FoodbornePathogensContaminants/Methylmercury/ucm191007.htm.

Food and Drug Administration. 2006. Mercury levels in commercial fish and shellfish. Accessed April 25, 2010, http://www.fda.gov/Food/FoodSafety/Product-SpecificInformation/Seafood/FoodbornePathogensContaminants/Methylmercury/ucm115644.htm.

Food and Drug Administration and Environmental Protection Agency. 2004. *What You Need to Know About Mercury in Fish And Shellfish: Advice For Women Who Might Become Pregnant, Women Who Are Pregnant, Nursing Mothers, and Young Children from the U.S. Food and Drug Administration and the U.S. Environmental Protection Agency.* Consumer brochure: EPA-823-R-04-005.

Formanek Jr., Raymond. 2001. Highlights of FDA food safety efforts: Fruit juice, mercury in fish. *FDA Consumer,* 35(2): 15–18.

Foulke, Judith. 1994. Mercury in fish: Cause for concern? *FDA Consumer,* 28(7):5–9.

Frickel, Scott. 2004. *Chemical Consequences: Environmental Mutagens, Scientist Activism and the Rise of Genetic Toxicology.* New Brunswick, NJ: Rutgers University Press.

Gerstenberger, Shawn, Adam Martinson, and Joanna L. Kramer. 2010. An evaluation of mercury concentrations in three brands of canned tuna. *Environmental Toxicology and Chemistry,* 29(2):237–242.

Got Mercury? and Environment Illinois. 2006. *Toxic Tuna: Mercury Contamination in Chicago Restaurant Tuna Sushi.* Chicago, IL: Got Mercury? and Environment Illinois Report.

Groth III, Edward. 2010. Ranking the contributions of commercial fish and shellfish varieties to mercury exposure in the United States: Implications for risk communication. *Environmental Research,* 110(3):226–236.

Hawthorne, Michael. 2006. New warning for canned tuna: Mercury risk for pregnant women too high, Consumer Reports says. *Chicago Tribune*, June 6.

Hightower, Jane. 2009. *Diagnosis: Mercury: Money, Politics, and Poison.* Washington: Island Press.

Hightower, Jane, and Dan Moore. 2003. Mercury levels in high-end consumers of fish. *Environmental Health Perspectives,* 111:604–608.

Hilts, Philip. 2004. *Protecting America's Health: The FDA, Business, and One Hundred Years of Regulation.* Charlottesville, NC: University of North Carolina Press.

Institute of Medicine. 1991. *Seafood Safety.* Washington, DC: National Academies Press.

Joyce, Kelly. 2006. From numbers to pictures: The development of magnetic resonance imaging and the visual turn in medicine. *Science as Culture,* 15(1):1–22.

Kraepiel, Anne, Klaus Keller, Henry Chin, Elizabeth Malcolm, and François Morel. 2003. Sources and variation of mercury in tuna. *Environmental Science & Technology,* 37(24): 5551–5558.

Lewis, Jack. 1985. The birth of EPA. Accessed May 19, 2010, http://www.epa.gov/history/topics/epa/15c.htm.

Lowenstein, Jacob, Joanna Burger, Christian W. Jeitner, George Amato, Sergios-Orestis Kolokotronis, and Michael Gochfeld. 2010. DNA barcodes reveal species-specific mercury levels in tuna sushi that pose a health risk to consumers. *Biology Letters,* October 23, 6(5): 692–695.

Malsch, Katherine, and Carroll Muffett. 2006. *Is Our Tuna Family Safe? Mercury in America's Favorite Fish.* Washington: Defenders of Wildlife.

Martosko, David. 2006. How big? *Washington Times*, October 22, B2.

Martosko, David. 2008. Fishy Omega-3 risks: Pendulum swings back to common sense. *Washington Times*, December 30, A17.

Mayer, Caroline, and Amy Joyce. 2005. The escalating obesity wars: Nonprofit's tactics, funding sources spark controversy. *Washington Post*, April 27, E1.

Mills, C. Wright. 1959. *The Sociological Imagination.* New York: Oxford University Press.

Mishima, Akio. 1992. *Bitter Sea: The Human Cost of Minamata Disease.* Translated by Richard Gage and Susan Murata. Toyko, Japan: Kosei Publishing Company.

Montague, Peter, and Katherine Montague. 1971. Mercury: How much are we eating? *Saturday Review,* (February 6):50–55.

Monterey Bay Aquarium. 2003. Seafood watch seafood report: Tunas volume VII yellowfin tuna. Accessed April 26, 2010, www.mbayaq.org/cr/cr_seafoodwatch/content/media/MBA_SeafoodWatch_YellowfinTunaReport.pdf.

Moore, Colleen. 2009. *Children and Pollution: Why Scientists Disagree.* New York: Oxford University Press.

National Fisheries Institute. 2010a. Healthy tuna: Tuna for all ages. Accessed May 28, 2010, http://www.healthytuna.com/health-nutrition/tuna-all-ages.

National Fisheries Institute. 2010b. Healthy tuna: Tuna facts. Accessed May 28, 2010, http://www.healthytuna.com/about-tuna/tuna-facts

National Oceanic and Atmospheric Association. 2002. Shrimp overtakes canned tuna as top U.S. seafood: Overall seafood consumption decreases in 2001. NOAA Press Release, August 28, NOAA 2002–113.

National Research Council (NRC). 2000. *Toxicological Effects of Methylmercury*. Washington: National Academies Press.

NOW. 2005. *Mercury in Our Fish*. Boston: Public Broadcasting Service, January 21.

Pianin, Eric. 2003. Federal warning on tuna planned: Mercury a danger to fetuses, children. *Washington Post*, December 11, A1.

Reardon, Jennifer. 2005. *Race to the Finish: Identity and Governance in an Age of Genomics*. Princeton, NJ: Princeton University Press.

Richwine, Lisa. 2004. Congress told FDA failed public on Vioxx. *Reuters*, November 8.

Roe, Sam, and Michael Hawthorne. 2005a. The mercury menace: How to minimize the risks of mercury. *Chicago Tribune*, December 11.

Roe, Sam, and Michael Hawthorne. 2005b. The mercury menace: Toxic risk on your plate. *Chicago Tribune*, December 11.

Saito, Hisashi. 2009. *Niigata Minamata Disease: Methyl Mercury Poisoning in Niigata, Japan*. Niigata City, Japan: Niigata Nippo Jigyosha.

Savitz, Jacqueline, Dr. Kimberly Warner, Melissa Anselmo, Simon Mahan, Beth Kemler, Dr. Michael Hirshfield, Dianne Saenz, and Dustin Cranor. 2008. *Hold the Mercury: How to Avoid Mercury When Buying Fish*. Washington: Oceana and Mercury Policy Project Report.

Steiner, Todd, and Michael Bender. 2010. *Comments on Special Supplemental Nutrition Program for Women, Infants and Children (WIC): Revision in the WIC Food Packages*. USDA. Docket ID: FNS-2006-0037-0003.

Sunderland, Elsie. 2007. Mercury exposure from domestic and imported estuarine and marine fish in the United States seafood market. *Environmental Health Perspectives,* 115(2):235–242.

Sunderland, Elsie, David P. Krabbenhoft, John Moreau, Sarah Strode, and William Landing. 2009. Mercury sources, distribution, and bioavailability in the North Pacific Ocean: Insights from data and models. *Global Biochemical Cycles,* 23:1–14.

US Food & Drug Administration Subcommittee on Science and Technology. 2007. *FDA Science and Mission at Risk: Report of the Subcommittee on Science and Technology*. Prepared for FDA Science Board.

US General Accounting Office. 2004. *Food Safety: FDA's Imported Seafood Safety Program Shows Some Progress, but Further Improvements Are Needed.*

US Geological Survey. 2010. Toxic Substances Hydrology Program. Accessed April 17, 2010, http://toxics.usgs.gov/definitions/biomagnification.html.

Waldman, Peter. 2005. Mercury and tuna: U.S. advice leaves lots of questions. *Wall Street Journal*, August 1.

Warner, Melanie. 2005. With sales plummeting, tuna strikes back. *New York Times*, August 19, C3.

Warrick, Joby. 1997. Senators press for report on mercury threat. *Washington Post*, May 22, A01.

# 6 How *USA Today* Constructs the Problem of Mercury Pollution

## *A Sociological Analysis of Risk and Blame*

*Christine Mowery and Sarah Jane Brubaker*

**CONTENTS**



> Much of the information the public needs to make decisions about environmental risk is not getting into newspapers.
>
> — **Peter Sandman, David Sachsman, Michael Greenberg, and Michael Gochfeld,** *Environmental Risk and the Press: An Exploratory Assessment*

## INTRODUCTION

In this chapter, we examine how one newspaper, *USA Today*, presents mercury pollution over a ten-year period, 2000–09. Specifically we focus on how the newspaper defines sources of mercury pollution as well as how it frames who is to blame and who is at risk.* In addition, the analysis considers which groups the news outlet uses as experts to help construct the social problem of mercury pollution. A better understanding of the framing of this risk can help assess public awareness and understanding of, and response to, mercury's hazards.

This study of how an individual newspaper depicts mercury pollution is embedded within a sociological framework. As a discipline, sociology is fundamentally interested in how the structure of a particular society influences an individual and group behavior. The structure of a society includes its culture, groups, statuses, roles, and social institutions (such as the family, religion, the economy, the polity, and the military). While each of these structural elements affects individuals within a society, the focus of this chapter is on the role of one institution, the mass media, and more specifically, one component of the mass media, the newspaper.† While social scientists interested in the mass media acknowledge its impact on almost every aspect of individuals' lives, from projecting and reinforcing a society's values to socializing boys and girls into "appropriate" gender roles, one of the key functions of the news media is as the primary source of information about local, national, and international events. Within this role, news outlets shape how individuals perceive these events and their roles within them.

Over the past three decades social scientists have become more and more interested in how the media construct the news, that is, how journalists select the information the public will see and hear as well as how news outlets frame this information, thereby defining the situation for the public. Thus, agenda-setting, or choosing news topics to focus on, is an important role of the mass media. Another factor to this agenda-setting, and of equal importance, is the frame in which the news is presented. A frame is an "interpretive schemata that simplifies and condenses 'the world out there' by selectively punctuating and encoding objects, situations, events, experiences, and sequences of actions within one's present or past environments" (Hunt, Benford, and Snow, 1994, 190). In other words a frame refers to the way the news outlets organize and present the issues they cover. This aspect of agenda-setting tells the public how to think about a particular issue. Which issues are covered and how, which sources are used, who is quoted and who is not (and thus, who is considered knowledgeable or an expert), all influence how an environmental hazard is presented to the general public. In this chapter, we analyze how *USA Today* frames the issue of mercury pollution.

---

* We chose to analyze *USA Today* because it has the highest circulation among a general audience, thus the "greatest reach" to the general population of the United States. We are interested in what information the average US citizen is getting from this particular form of print media.

† Social scientists divide the mass media into two types: print and electronic. Print media include books, newspapers, and magazines, while electronic media include the Internet, television, radio, and film. According to Marger (1993), to refer to these entities as "mass media" "implies that their communicative realm is extremely broad, often encompassing the entire society" (238).

The initial agenda-setting, choosing what topics to write about, is one aspect of how the media construct a social problem such as mercury pollution. Sociologists generally agree that social problems are socially constructed by a number of institutions in a society, meaning that a particular social problem such as mercury pollution, "does not become a social problem until (a) it is perceived to be so by a substantial number of individuals; and (b) there is a mobilization of resources directed toward the amelioration of this condition" (Faupel, Bailey, and Griffin, 1991, 294). Environmental issues become social problems through what people say and think about them, rather than by actual objective conditions that harm individuals or the environment.* The human element is thus highlighted at this stage of the agenda-setting because while sociologists view the media as a social institution, actual individuals within this institution choose what issues merit coverage and how these issues will be framed, thereby contributing to this social construction. Another human element then becomes how the public interprets the framing and decides how to respond. With respect to mercury pollution, the topics the media choose to cover shape the specific hazards about which the public thinks and becomes concerned. Viewing social problems as socially constructed in no way denies the harmful impact certain materials have on humans and the environment; rather, it merely highlights the role the media (as a social institution) have on how people begin to view an issue, as well as their response to it.

## LITERATURE REVIEW

In this section we provide a brief overview of sociological studies of how the media cover environmental risk. We also review how sociologists address how the media define who's at risk and who's to blame in terms of these risks, as well as which groups the media identify as sources of expert knowledge, or claims makers. These topics provide the framework for our analysis of how *USA Today* frames mercury pollution.

### Print Media and Coverage of Environmental Risk

While researchers have documented the impact the media have had on framing certain social problems, including violence and pornography (McCormack, 1978), missing children (Fritz and Atltheide, 1987), road rage (Best, 1991), and the potential harm done to children and society by heavy metal and rap music (Binder, 1993), only recently have social scientists turned their attention to the role of the media in constructing social problems related to environmental risk (see, for example, Zavestoski, Agenllo, Mignano, and Darroch, 2004; Shibley and Prosterman, 1998; Griffin and Dunwoody, 1997; Faupel, Bailey, and Griffin, 1991; Hansen, 1991; Brosius and Kepplinger 1990; Schoenfeld, Meier, and Griffin, 1990).

Assessments of media attention to environmental risks prior to the 1990s found that the media tended to offer little risk information from environmental contaminants

* Of course, news outlets are not the only players in the construction of an issue as an environmental risk. Other groups that have a stake in, and impact, how an issue is framed include environmental groups, various industries, regulatory agencies, and local, state, and federal governments.

to the general public (Ryan, Dunwoody, and Tanicard, 1991; Sandman, Sachsman, Greenberg, and Gochfeld, 1987; Singer and Endreny, 1987). For example, in a review article of risk reporting in the media, Kitzinger (1999) notes that the media do not necessarily prioritize the risks considered to be most important by scientists and epidemiologists; other factors, such as how unusual the risk is or risks that kill or injure many people at one time, play a role in the amount of coverage the media give to a particular hazard.

## Print Media and Who's to Blame and Who's at Risk

Despite this criticism of overall lack of coverage, social scientists examine the news coverage of environmental hazards, particularly in individual case study form. In a handful of these studies researchers have explored the issue of who's at risk and who's to blame for various environmental hazards. These analyses highlight the media's role in targeting particular groups for responsibility or culpability when damage is done to either individuals or the environment. Cheek's (1997) analysis of Australian print-based media's representation of toxic shock syndrome (TSS) between 1979 and 1995, for example, illustrates that TSS was framed as an issue of "neglect of care for the self, and of carelessness on the part of individual women. ... The threat of TSS is represented as lying not in the syndrome itself, but in ignoring the steps to be taken to avoid it" (196). Thus, individual women are portrayed as being responsible for their own health and in turn at fault if they do contract TSS, with no blame placed on tampon manufacturers. Cheek goes on to note that it is in the interest of powerful societal groups such as manufacturers of tampons that "individual women and their use of tampons, rather than tampons themselves, be framed as the 'problem' to be addressed" (198).

In their study of childhood lead poisoning as a public health problem, Shibley and Prosterman (1998) assert that, although poor and minority children are at greater risk of lead poisoning, print news media are "interested in what will appeal to readers. Newspapers with a white, middle-class audience will tend to present lead poisoning as a mainstream (i.e., epidemic) problem, as an issue directly relevant to the concerns of those who read the paper" (54). Thus, in part to appeal to their audience, the newspapers under study downplayed the fact that risk of lead poisoning is unevenly distributed across the population. The framing of lead poisoning as an epidemic impacting all children was also influenced by media reliance on certain claims makers.

## Print Media and Reliance on Claims Makers

Another aspect of frame-making relates to who news sources choose as "claims makers." Claims makers such as social activists, industry stakeholders, public health officials (such as those from the Centers for Disease Control), and professional experts help shape how an environmental risk is framed for the public (Kitzinger, 1999; Stallings, 1990; Singer and Endreny, 1989). News reporters often rely on claims makers' actions, press releases, and press conferences when constructing a particular definition of environmental risk, as well as acceptance of a particular version of risk. Uncovering the choice of claims makers and what organizations they represent is important since very different messages concerning an environmental hazard can come from different sources. As Snow and Benford (1988) point out, for example, social activists and their movements' organizations are continually framing social issues "in ways that are intended to mobilize potential adherents and constituents, to garner bystander support, and to demobilize antagonists" (198). In their analysis of the framing of childhood lead poisoning as an environmental health problem, for example, Shibley and Prosterman (1998) identify public health officials, medical "experts," industry interests (such as the real estate industry and insurance companies), and community advocates as the major players in defining lead issues in the national media. They conclude that even though some children (i.e., minority and lower-class children) are at much greater risk for lead poisoning than others, lead advocates and the environmental justice movement frame the problem as an epidemic in which all children are at risk so as not to undermine broad public support and future allocation of public resources for addressing lead poisoning issues.

Despite efforts to shape discussion of an issue in a particular way, social movement activists are seldom included as experts in news stories. For example, in their analysis of television coverage of environmental risk stories presented on ABC, CBS, and NBC evening news broadcasts from January 1984 to February 1986, Greenberg et al. (1989) found that the government accounted for 28% of sources, industry for 13.2%, and advocacy groups a mere 6.8%. In their study of how the local newspapers framed hazardous waste as a social problem in Sumter County, Alabama, Faupel et al. (1991) suggest that the tone of reporting, either positive or negative toward the local hazardous waste facility, was directly related to the media reliance on either industry or regulatory agency spokespeople. And finally Zavestoski et al. (2004) argue that the lack of community mobilization or opposition to dioxin contamination in Woonasquatucket River in Rhode Island was due in part to the influence of the Environmental Protection Agency's (EPA) shaping of the media framing of this toxic event, in particular the EPA argument that scientific uncertainty existed regarding dioxin's human health effects and the EPA downplaying the public health risk of dioxin.

The literature addressing media framing of environmental risk provides a basis for our analysis of how *USA Today* frames mercury pollution. In particular, our guiding research questions were: How is mercury pollution defined and presented? Who is identified as at risk in these articles on mercury pollution? Who is identified as to blame for mercury pollution? And whom do the journalists rely on as claims makers?

## METHODOLOGY

Sociologists commonly use content analysis when investigating the material used in the mass media (including magazines, newspapers, television shows, movies, children's books, and popular music).* At its most basic, content analysis is the

* In our study we focused on print media. Another option was to focus on nonprint media, such as television, radio, or some type of Internet media, but it is often difficult to set clear parameters around nonprint media and its content can change quickly.

systematic coding and objective recording of data (the data being written or recorded communications, such as song lyrics, newspaper articles, and television commercials) in order to understand the messages communicated through various types of texts. In bringing to light underlying themes or regular patterns hidden within texts, content analysis helps researchers explore the messages being consumed by millions of people who read newspapers, watch television, listen to music, and so on. Additionally researchers are able to explore how the media make certain issues salient while completely ignoring others, and how this can in turn influence public beliefs and opinions regarding these issues (Riffe, Lacy, and Fico, 2005).

Through content analysis we can also explore questions involving which claims makers are used by the media when they are writing about a particular issue. Are the experts typically scientists and if so, who are their employers? Is the information consistently coming from the same group or similar groups of sources/experts? Content analysis also helps to answer questions such as: Are the issues framed by the media in ways that favor the economic and politically elite? Who is implied to be at risk, and who is culpable? How do the media portray the issue—as problematic, as natural, as a hazard, caused by human practices? And, ultimately, how do the media shape the reality of the issue?

In our exploration of newspaper framing of mercury pollution we were interested in analyzing the print media that are read by the most people from a general audience. That is, we were not interested in specialized media such as academic journals or trade magazines, or newspapers specific to a particular region. We did, however, decide to focus only on the United States. Using the Audit Bureau of Circulations (accessabc.com) we identified *USA Today* as the newspaper with the highest circulation among a general audience. We then needed to set parameters in terms of time, i.e., how many months or years of issues to analyze. Using a common, readily-available search engine called LexisNexis we conducted a search for the past ten years (i.e., 2000–09) of all articles that included the terms mercury and either pollution, environment, or risk. (We discovered that using the single word "mercury" as a search term often turned up articles about cars, since that is the name of a major automobile manufacturer.) Our search resulted in 140 articles from January 2000 through December 2009 that addressed mercury pollution or risk. We analyzed the entire population of articles, that is, every article published by *USA Today* in the ten-year timeframe that addressed mercury pollution.

Content analysis is essentially a coding operation whereby the oral or written communications under study are systematically classified according to some conceptual framework. Researchers develop reliable coding categories and then apply these categories to the data under study. Coding includes both quantitative and qualitative approaches. For example, in terms of quantitative coding we could count how many times we see certain words or phrases (fish, mining, etc.) in our sample of newspaper articles. These numbers can then be shown with graphs or tables, as we do below. Alternatively we could tally the number of column inches assigned a particular coding category or the number of column inches assigned to articles written about mercury pollution. Regardless of the specific categories utilized, coding transforms qualitative content into quantitative data (Neuman, 2003). Qualitative coding involves identifying, comparing, and describing themes and ideas that various media convey, and how they construct and communicate the themes through certain words, images, and other signifiers.

After we had collected the 140 articles we began coding. Our unit of analysis was the individual article. Through the coding process we coded, or labeled, each article in terms of the following themes: the type of pollution or risk that was being addressed, the groups identified as most at risk from mercury pollution or poisoning, and the groups identified as most responsible or to blame for the risk. We also coded the claims makers that the article identified. We counted the instances of each of these categories as part of our quantitative analysis and then identified and described the major themes of each as part of our qualitative analysis.

One challenge to conducting this analysis is that some articles address multiple issues related to mercury. For example, some discussed both coal-burning power plants and how the mercury pollution in the air goes into the ground, the water, and eventually into the fish that we eat (thus, highlighting the dangers of mercury pollution via coal-burning power plants and consumption of mercury-laden fish). In these instances we coded the article as addressing both types of mercury risk, which means that the total number of types (or sources) of mercury pollution exceeds the total number of articles. Another challenge was that not every article addressed each theme we were interested in. For example, not all articles identified specific claims makers or specific groups at risk or to blame for mercury pollution. Some articles were focused, for example, on environmental policy in general and discussed several different environmental problems including mercury, but these articles only gave mercury pollution cursory attention. Although one approach to content analysis is to assess the amount of space given to a topic (e.g., number of words or sentences devoted to it) we did not find it necessary to analyze that level of detail. We did code for each of the issues we were focused on when they were available. We discuss the results of our content analysis below.

## RESULTS

### The Quantity of Coverage

Overall, 140 articles in ten years of the newspaper with the highest circulation in the country, *USA Today*, are not many, suggesting that mercury pollution has not been defined as a major issue by the print media. In comparison, for example, Shibley and Prosterman (1998) found 217 articles on lead poisoning for a mere four-year time period.* In Table 6.1 we indicate the number of *USA Today* articles on mercury pollution that were published in each year of our timeframe. We also indicate the primary source of mercury pollution discussed within each article. For this table we coded each article based on the primary source of pollution addressed rather than coding all sources mentioned in the article.

In Figure 6.1 we show all sources of mercury pollution addressed in these articles in this time period in graphic form. The highest number of articles on mercury

* Their analysis was derived from the following three newspapers: *The New York Times*, *Chicago Tribune*, and the *Chicago Defender*.

**TABLE 6.1**
**Number of Articles on Mercury Pollution Per Year by Primary Source**

| Year | Fish Consumption | Coal-Fired Power Plants | Electronics Disposal | Vaccines | Other | General Pollution | Total |
|---|---|---|---|---|---|---|---|
| 2000 | 0 | 3 | 1 | 1 | 8 | 1 | 14 |
| 2001 | 0 | 8 | 0 | 1 | 1 | 2 | 12 |
| 2002 | 8 | 3 | 2 | 2 | 4 | 0 | 19 |
| 2003 | 5 | 2 | 0 | 1 | 1 | 1 | 10 |
| 2004 | 13 | 2 | 2 | 4 | 0 | 2 | 23 |
| 2005 | 5 | 3 | 0 | 2 | 0 | 1 | 11 |
| 2006 | 5 | 0 | 2 | 2 | 1 | 3 | 13 |
| 2007 | 3 | 4 | 0 | 1 | 6 | 0 | 14 |
| 2008 | 1 | 1 | 2 | 0 | 5 | 4 | 13 |
| 2009 | 2 | 1 | 1 | 2 | 4 | 1 | 11 |
| Total | 42 | 27 | 10 | 16 | 30 | 15 | 140 |

pollution were published in *USA Today* in 2004 and the lowest number in 2003. The high number of articles in 2004, twenty-three, likely reflects the 2004 joint fish consumption advisory issued by the EPA and the FDA. (Chapter 5 discusses the 2004 federal fish advisory.) Overall, fish consumption was the primary type of pollution addressed across the ten-year period, and the greatest number of articles addressing fish consumption were published in 2004. The first time that fish consumption was a primary theme of an article on mercury was in 2002, the same year that two major medical studies were published that addressed this issue.

Although other mercury-related issues constituted the second-highest category of themes, it included a number of different topics that received minimal attention from *USA Today* during this time period. These issues included concern about exposure to and removal/disposal of various items that include mercury such as glass thermometers, other types of medical equipment such as blood pressure gauges, home gas meters, dental fillings, and new high-efficiency lightbulbs.

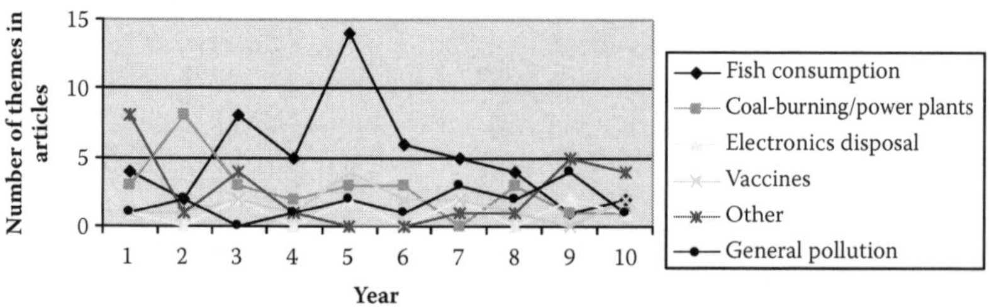


**FIGURE 6.1** Number of each primary source of mercury pollution addressed in articles on mercury pollution by year, 2000–2009.

The next most prevalent source of mercury pollution discussed in our group of articles was coal-burning power plants. Articles addressing this theme focus primarily on how plants expel mercury pollution into the air. Although some articles fail to elaborate on how the pollution poses risk or can be harmful, others explain how pollution transfers from the air, to the ground water, and then is ingested by the fish that we eat.

General pollution issues or environmental policy were the topics of fifteen articles in this time period, and each mentioned mercury as a pollution problem. Most did not elaborate on any particular aspect of mercury, however. Similarly, sixteen articles addressed potential risks and side effects of vaccinations, mentioning the historical use of thimerosal, a derivative of mercury, which is no longer used in vaccinations in the United States; however, it is still commonly used as a preservative in other parts of the world (NIAID, 2009). In the cases of both general pollution and vaccinations mercury was only mentioned briefly as one example of the issue without going into detail. For this reason we do not provide an analysis of these topics.

Electronics disposal constituted the least mentioned source of mercury pollution in the articles, but the coverage was more in depth with this issue than with general pollution or vaccinations. Specifically these articles discussed how electronics such as computer monitors contain mercury, and when not disposed of or recycled properly, the mercury can leak into the ground. Some articles elaborated that the mercury can eventually enter the water, polluting the fish that we eat.

## The Quality of Coverage

In this section we discuss the qualitative findings of our content analysis, based on our guiding questions: How is mercury pollution defined and presented? Who is identified as at risk in these articles on mercury pollution? Who is identified as to blame for mercury pollution? And whom do the journalists rely on as claims makers? It is important to keep in mind that the media coverage of this issue, like that of related environmental issues, is not clear-cut or unequivocal. That is, even this one media source, *USA Today*, presents conflicting and contradictory information across individual articles. This conflicting information makes it difficult for readers to determine the truth about mercury risks. In our discussion of these issues we identify the main themes and the extent to which the media present conflicting arguments or views.

### Fish Consumption

The articles we analyzed focusing on fish consumption addressed mercury contamination of fish that many people eat. Some articles indicated the types and sources of fish that are most likely to be contaminated by mercury but some did not. Some articles emphasized that the main types of fish that have been tested and found to contain dangerous levels of mercury live in lakes and streams and are caught by recreational or sports fishermen. These articles emphasize that the fish people buy from supermarkets or eat in restaurants are not vulnerable to mercury poisoning. One such article, for example, is titled "Most people aren't in danger" (Weiss 2004) because of these factors. Other articles, however, state that canned tuna, which is available in restaurants and at supermarkets, can contain dangerous levels of mercury. Similarly some articles address food activists' efforts to label with mercury warnings fish sold in supermarkets, suggesting that commercially caught and sold fish can be tainted in addition to those caught recreationally in lakes and streams. This quotation from a 2004 article illustrates this point:

> While EPA Administrator Michael Leavitt points out that tuna and other fish bought in stores are not at issue in the recent advisories, he fails to note that his agency joined the U.S. Food and Drug Administration in warning women and children to limit how much tuna they eat because of mercury contamination. (2004, 16A)

Most of the articles on fish as a source of mercury poisoning identified pregnant women and women of childbearing age as most at risk of mercury poisoning because of the potential neurological damage to the fetus if the mothers consume mercury. Other articles identified children as at risk. Only a few suggested that all people who eat fish are potentially at risk.

Claims makers in these articles included medical doctors, medical researchers, environmental activists, and representatives of governmental agencies including the EPA and the FDA. Although not all articles explicitly addressed blame or responsibility for mercury pollution, those that did focused on these few major groups. Some noted that the EPA and FDA are charged with maintaining health standards, suggesting that they are to blame when standards are either too loose or not sufficiently enforced. Some articles blamed manufacturers (e.g., of electronics that are improperly disposed of) and industries (such as power plants that burn coal) for polluting the air, the water, and ultimately the fish that people eat. A few articles suggested that exposure to mercury is the consumers' responsibility for eating the fish that is most likely to be contaminated with mercury. In particular pregnant women are often admonished to limit their consumption of fish, suggesting that they are responsible for exposing fetuses to mercury. For example, a 2002 article stated,

> "Tuna is the most consumed fish in the country," Bender says. "If you're a pregnant woman and you eat over two cans of tuna per week, you can go over" safe levels of mercury. The FDA currently recommends that women who are or could become pregnant limit all fish to 12 ounces a week. (Manning, 2002, 1A)

### Coal-Burning Power Plants

Articles that target coal-burning power plants as a source of mercury pollution (the second largest group of articles in our study) focus primarily on how plants expel mercury pollution into the air. For example, an article from 2004 stated the following:

> Mercury is emitted primarily by incinerators and power plants that burn coal. The EPA plans to publish rules restricting mercury from power plants by mid-2005, although environmentalists say the preliminary draft of those rules does not go far enough. (Weise and Watson, 2004, 1A)

Although some articles fail to elaborate on how the pollution poses risk or can be harmful, others explain how pollution transfers from the air, to the groundwater, and then is ingested by the fish that we eat. It is unclear from these articles whether humans are harmed by breathing the polluted air directly, though several articles identify a general neurological risk or potential harm to the nervous system without identifying any particular groups that are at greater risk than others. The articles that link coal burning to water pollution and fish contamination also identify pregnant women and fetuses as at risk.

In terms of blame, the articles that address this issue target either the owners of the factories for burning the coal or the EPA for not setting or enforcing stricter emission standards. For example, one article stated that, "Parts of the answer, such as mercury-emission controls and better management of fish stocks, are in the laps of governments" (Weise, 2005, 1D).

### Electronics Disposal

The third most prevalent source of mercury among the articles we found was electronics disposal and recycling, sometimes referred to as "E-waste recycling." Most of the articles discussed how mercury is present in computer monitors and, when disposed of improperly rather than recycled safely, can leak into the ground and eventually pollute the water. The articles on this issue identified numerous groups that could be to blame for this environmental hazard, including manufacturers of electronics for not taking responsibility for their disposal, the EPA for not setting or enforcing higher standards or putting into place more accessible electronics recycling options, and even consumers for carelessly disposing of electronics. A 2004 article explained the problem thus:

> [C]omputer and electronics makers are increasingly pressured to recycle. Most electronics contain lead, mercury or other substances that can be harmful if dumped in landfills. Recycling solves that problem. But it's costly: about $20 per PC. ... But electronics makers may soon have no choice. California and Maine recently passed laws requiring manufacturers to take some responsibility for recycling. (Kessler 2004, 4B)

These articles were much more likely to focus on blame than risk, with only one article identifying brain damage as a specific type of potential risk without specifying which groups were at risk for this type of harm.

## Spreading the Blame

In addition to blaming US government agencies, manufacturers, or consumers for mercury pollution and risk some articles blame other countries' environmental practices. For example, one article stated:

> And though mercury emissions in the USA are going down, there's no guarantee that mercury pollution in lakes and rivers will do the same. That's because mercury travels into the USA from around the world. China and other East Asian nations are especially big sources, as is Western Europe. More than 50% of the mercury pollution in the USA comes from outside the country, Leavitt said. (Watson, 2004, 3A)

The articles that blamed other countries for mercury pollution typically identified Asian countries, including China and India, as the main culprits. Here is another example:

> Mercury from China, dust from Africa, smog from Mexico — all of it drifts freely across U.S. borders and contaminates the air millions of Americans breathe, according to recent research from Harvard University, the University of Washington and many other institutions where scientists are studying air pollution. There are no boundaries in the sky to stop such pollution, no Border Patrol agents to capture it. (Watson 2005, p. 1A)

At the same time that some articles cite other countries' contribution to the US's pollution problems, one article suggested that the United States. may also be contributing to theirs:

> That's one reason BAN and other activists have ramped up efforts to slow the unfettered export of the USA's e-waste to poorer countries. There are signs they're getting results. Since August, when the Government Accountability Office released a blistering investigative report declaring that exported U.S. e-waste was often disposed of unsafely in countries such as China and India, BAN has received pledges from dozens of electronics recyclers that they won't export. (Schmit 2008, 1B)

Overall when blame was addressed there was relatively equal distribution across governmental agencies, industry, and consumers. As we discuss later in this chapter this range and variation, coupled with the emphasis on international blame, results in a fragmented framing of the problem that does not provide a comprehensive or unified perspective.

## Claims Makers

Across the various types of mercury pollution sources, the same general groups were identified as claims makers, i.e., governmental agencies such as the FDA and EPA, industry, medical researchers, and environmentalists. Table 6.2 provides an overview of the quantitative results of our analysis of claims makers. The EPA was the most prevalent, and although the FDA was the least mentioned, taken together as governmental agencies they represent a category significantly larger than any other. It is important to recognize, however, that groups such as scientists and researchers can be employed by

**TABLE 6.2**
**Claims Makers Identified in *USA Today* Articles, 2000–09**

| Claims Makers | Number of Articles |
|---|---|
| Environmental Protection Agency (EPA) | 53 |
| Scientists, academics, or academic studies/researchers | 43 |
| Environmental activists/advocates or consumer groups | 39 |
| Medical/health experts, providers, or agencies (e.g., Centers for Disease Control or National Institutes of Health) | 28 |
| Industry representatives/spokespeople (e.g., electrical power plants or electronics manufacturers) | 25 |
| Food and Drug Administration (FDA) | 19 |

and represent governmental agencies, medical organizations, or even industries, which makes it difficult to clearly identify the claims makers around this issue.

Although the knowledge and authority of government agencies and medical researchers might be used by those on either side of the issue, industries tended to resist environmental controls and regulation and downplay risk and harm, while environmentalists tended to support stricter regulations and emphasize greater risk. For example, environmental activists are more likely to present an anti-industry position, and industry spokespeople are more likely to blame consumers or government. One article helped to illustrate this issue by presenting various perspectives:

> Confusing messages and inflated fears are keeping the public from getting the benefits that come from eating fish, says Dariush Mozaffarian, lead author of a report by the Harvard School of Public Health. "We were surprised at how little evidence there actually is for some of these harms." ... Findings are published in the Journal of the American Medical Association. ... [On the other hand] public concerns about the risks are "overblown," based on "fear and misinformation," says William Hogarth of the National Oceanic and Atmospheric Administration. NOAA and the Food and Drug Administration funded the study. Gerald Leape of the National Environmental Trust says the risks posed by contaminants are greater than the studies suggest. "There's growing evidence that increased levels of fire retardants and PCBs found in fish can cause severe developmental and neurological problems," he says. (Weise, 2006, A1)

Although the last quotation in the passage above is from an environmentalist, this group was not the most prevalent type of claims makers identified in the articles. The following quote represents the most extensive attention given to the environmentalist perspective as claims maker:

> "The big lesson here is the Bush administration is happy to move forward with environmental controls, but only to the extent industry is comfortable with them," said Frank O'Donnell of Clean Air Watch, a non-profit environmental group. But the EPA says the mercury rule went as far as it could without excessive costs to companies and their customers. The rule "will virtually eliminate elevated levels of mercury American women and children may face solely because of U.S. power plants," said Jeffrey Holmstead, the EPA's air director. What the two rules will do: The first will lower permitted levels of smog-forming chemicals and particle pollution from power plants in 28 Eastern states. Both have been linked to heart and lung problems. The second limits mercury from plants across the nation. Mercury settles into waterways and makes fish unsafe to eat. Children whose mothers were exposed to mercury when they were pregnant can have developmental problems. A power plant can meet both limits by installing pollution-control technology. But a plant can also satisfy the requirements by paying another plant to reduce emissions further to offset a smaller reduction by the first plant. That is known as trading pollution credits. Environmental groups, including Environmental Defense and the U.S. Public Interest Research Group, object to mercury trading. They say it could allow mercury to accumulate in "hot spots" near plants that decide not to cut emissions. (Watson, 2005, 4A)

As with the issue of blame, the variation in the media's use of claims makers, and the relatively comparable use of each across the articles, also contributes to a fragmented framing of the problem, leaving the general public with no clear understanding of the issue. It is almost impossible to determine where the majority of scientists agree or how sound the overall scientific research is given the news' emphasis on sound bites. In an attempt to present a balanced representation of opposing views, the news media steer away from providing a strong or clear message to the public.

## DISCUSSION AND RECOMMENDATIONS FOR FUTURE RESEARCH

In this chapter we analyze how *USA Today* frames the environmental hazard of mercury pollution, specifically what is defined as mercury pollution as well as who is to blame for this environmental problem and who is at risk. Additionally we are interested in understanding what groups the media rely on when constructing this social problem.

Through content analysis we discovered that there has in fact been very little attention paid to mercury pollution from the mass media, or more specifically from *USA Today*. What attention has been paid appears from our analyses to be unfocused, targeting various sources of mercury pollution as well as a variety of responsible parties, from owners of coal-burning factories to manufacturers of electronics to US government agencies for not regulating mercury emissions. That the media present this as such a multifaceted problem in such a fragmented way may be the reason that no single overarching frame could be discovered through our content analysis. Given the multitude of sources of mercury pollution, as well as questions as to who exactly is to blame and who is at risk from these sources, this newspaper has not developed any comprehensive frame for this environmental hazard with regards to what exactly the public should be concerned about.

Our analyses also revealed that only a few claims makers were identified in these articles, and we suggest that the lack of a comprehensive frame or sense of urgency regarding mercury pollution may be linked to the lack of social movement activists as claims makers in the majority of the articles. One would expect that environmental groups concerned with mercury pollution would attempt to frame the problem as an immediate danger, but since most "experts" cited in the article were government agencies it is not surprising that mercury pollution was not framed as an environmental hazard needing immediate action.

For sociologists it is always important to consider the demographic groups affected by any particular social problem and the ways in which groups are affected differently. This issue brings us back to the human element and how mercury affects groups of people. Although we did not code specifically for this element our analyses of risk and blame do allow us to infer some of the gender and social class effects of how mercury pollution is framed. In terms of gender, the group most identified as at risk is pregnant women. This framing places value on women in terms of their role as mothers but also places blame on them by holding them responsible for protecting their fetuses from harm. We can speculate that owners of industry and sports fishermen tend to be men, but none of the articles bring particular attention to this gendered issue.

A number of social class issues emerge from our analysis as well. One article identified the following occupations as putting people at risk for mercury poisoning, "Dentists, gold extractors, chemical plant workers, miners and others may come into contact. Gastrointestinal problems, tremors, central- and nervous-system effects can result" (Armour, 2000, 1A). Although dentists are clearly from higher social class positions, dental assistants (who would be considered middle class) typically spend more time with the patients, and the other occupations represent working-class arenas. Similarly recreationally caught fish and canned tuna are more likely to be consumed by people from working and lower classes than by those from higher social classes. Only one article (Bacon, 2006) identified sushi as a potential source of mercury, which is more expensive and more likely to be consumed by those from higher social classes. Finally the practice of exporting our e-waste to poor countries suggests that people in those countries are at greater risk. Overall, although not explicitly discussed in the articles, the consequences of mercury poisoning and pollution do not appear to be the same for all groups of people. More research should be done to better understand the differential effects of mercury on women and men, rich and poor, and other categories of people, and ultimately more effort should be made by the media to communicate those various risks to the public.

Another area that is important to investigate is why the media have not focused more on this issue. Despite the medical reports in 2002 that identified the risk of mercury poisoning from consuming fish caught in contaminated water, the concern communicated through the media was limited and short-lived. It seems that because of the complexity of the issue, the various perspectives on who is at risk and who is to blame, and the conflicting arguments of various claims makers the media have not constructed an overall frame of the issue to communicate to the public. Future research should investigate this issue further, including an examination of how changes in the political climate (such as Obama's presidency as compared to Bush's) contribute to changes in attention to environmental issues.

Overall, content analysis of media coverage and framing of mercury pollution enables us to understand how environmental risk is socially constructed and communicated to the general public. It helps us to identify the claims makers, those experts whose authority and knowledge we rely on to understand the world around us and how our environment affects us, as well as how the media place social members into significant groups with specific and often competing interests with respect to environmental factors. We hope to see more researchers engage in this type of analysis so that ultimately the public can have access to more accurate information and engage in more informed decision-making when it comes to environmental issues such as mercury pollution. Unfortunately our analysis suggests that twenty years after Sandman et al.'s observation that opened our chapter, i.e., that newspapers fail to provide the public with sufficient information about environmental risks, it still holds true today at least with respect to mercury pollution.

## REFERENCES

Armour, Stephanie. 2000. Workers unwittingly take home toxins: Employees endanger their loved ones when invisible but poisonous substances cling to their belongings. *USA Today* news section, October 5.

Bacon, John. 2006. Sheehan arrested as she protests in NYC. *USA Today* news section, March 7.

Best, Joel. 1991. "Road warriors" on "hair trigger highways": Cultural resources and the media's construction of the 1987 freeway shootings problem. *Sociological Inquiry,* 61:327–345.

Binder, Amy. 1993. Constructing racial rhetoric: Media depictions of harm in heavy-metal and rap music. *American Sociological Review,* 58:753–767.

Brosius, Hans-Bernd, and Hans Kepplinger. 1990. The agenda-setting function of television news. *Communication Research,* 17:183–211.

Cheek, Julianne. 1997. (Con)textualizing toxic shock syndrome: Selected media representations of the emergence of a health phenomenon 1979–1995. *Health,* 1:183–203.

Coltrane, Scott, and Michele Adams. 1997. Work–family imagery and gender stereotypes: Television and the reproduction of difference. *Journal of Vocational Behavior,* 50:32–47.

Curb mercury emissions for health of mothers, children. 2004. *USA Today* news section, September 2.

Evans, Lorraine, and Kimberly Davies. 2000. No sissy boys here: A content analysis of the representation of masculinity in elementary school reading textbooks. *Sex Roles,* 42:255–270.

Faupel, Charles, Conner Bailey, and Gary Griffin. 1991. Local media roles in defining hazardous waste as a social problem: The case of Sumter County, Alabama. *Sociological Spectrum,* 11:293–319.

Fritz, Noah, and David Altheide. 1987. The mass media and the social construction of the missing children problem. *Sociological Quarterly,* 28:473–492.

Gamson, William, and Andre Modigliani. 1989. Media discourse and public opinion on nuclear power: A constructionist approach. *American Journal of Sociology,* 95:1–37.

Gitlin, Todd. 1979. Prime time ideology: The hegemonic process in television entertainment. *Social Problems,* 26:251–266.

Glascock, Jack. 2001. Gender roles on prime-time network television: Demographics and behaviors. *Journal of Broadcasting and Electronic Media,* 45:656–669.

Greenberg, Michael, David Sachsman, Peter Sandman, and Kandice Salomone. 1989. Network evening news coverage of environmental risk. *Risk Analysis,* 9:119–126.

Griffin, Robert, and Sharon Dunwoody. 1997. Community structure and science framing of news about local environmental risks. *Science Communication,* 18:362–384.

Hansen, Anders. 1991. The media and the social construction of the environment. *Media, Culture, and Society,* 13:443–458.

Hunt, Scott, Robert Benford, and David Snow. 1994. Identity fields: Framing processes and the social construction of movement identities. In *New Social Movements, from Ideology to Identity*, Enrique Larana et al., eds., 185–208. Philadelphia: Temple University Press.

Kessler, Michelle. 2004. Office Depot will recycle old PCs free in "fabulous" offer. *USA Today* money section, July 13.

Kitzinger, Jenny. 1999. Researching risk and the media. *Health, Risk & Society,* 1:55–69.

Kuypers, Jim. 2002. *Press Bias and Politics: How the Media Frame Controversial Issues.* Westport, CT: Praeger.

Kuypers, Jim. 2009. *Bush's War: Media Bias and Justifications for War in a Terrorist Age.* Lanham, MD: Rowman & Littlefield Publishers Inc.

Manning, Anita. 2002. People who eat a lot of fish may run health risk. *USA Today* news section, November 5.

Marger, Martin. 1993. The mass media as a power institution. In *Power in Modern Societies,* Marvin Olsen and Martin Marger, eds., 238–249. Boulder, CO: Westview Press.

McCormack, Thelma. 1978. Machismo in media research: A critical view of research on violence and pornography. *Social Problems,* 25:544–555.

National Institute of Allergy and Infectious Diseases. 2009. Vaccines. U.S. Department of Health and Human Services, National Institutes of Health. Accessed May 23, 2010, http://www.niaid.nih.gov/topics/vaccines/research/pages/vaccines.aspx.

Neuman, William Lawrence. 2003. *Social Research Methods: Qualitative and Quantitative Approaches*, 5th ed. New York: Pearson.

Peterson, Sharyl Bender, and Mary Alyce Lach. 1990. Gender stereotypes in children's books: Their prevalence and influence in cognitive and affective development. *Gender and Education,* 2:185–197.

Riffe, Daniel, Stephen Lacy, and Frederick Fico. 2005. *Analyzing Media Messages: Using Quantitative Content Analysis in Research*, 2nd ed. Mahwah, NJ: Lawrence Erlbaum Associates.

Ryan, Michael, Sharon Dunwoody, and James Tanicard. 1991. Risk information for public consumption: Print media coverage of two risky situations. *Health Education Quarterly,* 18:375–390.

Sandman, Peter, David Sachsman, Michael Greenberg, and Michael Gochfeld. 1987. *Environmental Risk and the Press: An Exploratory Assessment.* New Brunswick, NJ: Transaction Books.

Scheufele, Dietram. 1999. Framing as a theory of media effects. *Journal of Communication,* 29:103–123.

Schmit, Julie. 2008. Watchdogs hound e-waste exports: USA's trashed TVs, computer monitors can make a toxic mess. *USA Today* money section, December 30.

Schoenfeld, A. Clay, Robert Meier, and Robert Griffin. 1990. Constructing a social problem: The press and the environment. *Social Problems,* 27:38–61.

Shibley, Mark, and Annette Prosterman. 1998. Silent epidemic, environmental injustice, or exaggerated concern? Competing frames in the media definition of childhood lead poisoning as a public health problem. *Organization and Environment,* 11:33–58.

Shih, Tsung-Jen, Rosalyn Wijaya and Dominique Brossard. 2008. Media coverage of public health epidemics: Linking framing and issue attention cycle toward an integrated theory of print news coverage of epidemics. *Mass Communication & Society,* 11:141–160.

Singer, Eleanor, and Phyllis Endreny. 1987. Reporting hazards: Their benefits and costs. *Journal of Communication,* 37:10–26.

Snow, David, and Robert Benford. 1988. Ideology, frame resonance, and participant mobilization. In *From Structure to Action: Social Movement Participation Across Cultures*, Bert Klandermans, Hans Kriesi, and Sidney Tarrow, eds., 197–217. Greenwich, CT: JAI Press.

Stallings, Robert. 1990. Media discourse and the social construction of risk. *Social Problems,* 37:80–95.

Thompson, Teresa, and Eugenia Zerbinos. 1995. Gender roles in animated cartoons: Has the picture changed in 20 years? *Sex Roles,* 32:651–673.

Watson, Traci. 2004. States look harder for mercury. *USA Today* news section, August 25.

Watson, Traci. 2005. Air pollution from other countries drifts into USA. *USA Today*, news section, March 14.

Watson, Traci. 2005. Groups condemn EPA's mercury rule. *USA Today* news section, March 16.

Weise, Elizabeth. 2004. Most people aren't in danger. *USA Today* news section, August 25.

Weise, Elizabeth. 2005. Eating fish: There's a catch. *USA Today* life section, October 26.

Weise, Elizabeth. 2006. More benefits than risks in higher fish consumption. *USA Today* news section, October 18.

Weise, Elizabeth, and Traci Watson. 2004. Warnings on river, lake fish jump. *USA Today* news section, August 25.

Zavestoski, Stephen, Kate Agnello, Frank Mignano, and Francine Darroch. 2004. Issue framing and citizen apathy toward local environmental contamination. *Sociological Forum,* 19:255–283.

# 7 Input/Output
## *Researching and Communicating Mercury Issues Online*

*John Garnett Drummond*

**CONTENTS**

> Where is the wisdom we have lost in knowledge?
> Where is the knowledge we have lost in information?
>
> **—T.S. Eliot,** *The Rock*

## INTRODUCTION

If a person Googles "mercury pollution," the popular search engine identifies 392,000 items in 0.35 seconds. How can a person know what information is reliable or not? What are the best ways to transmit "good" information in a circus of digital media? This chapter is a collection of ideas and advice surrounding online communications about mercury. In many ways it's more about the communication itself—the ways readers read it, the ways authors communicate it, and methods online researchers can use to investigate where these communications are coming from, and who and where its authors are. The origin for this chapter came about from a class I taught in which students learned to evaluate Internet sources by looking at mercury-related online resources.

## INPUT: RHETORICAL AND TECHNICAL TOOLS

When teaching university students techniques to evaluate information online I recommend the cyclical use of a compact arsenal of rhetorical and technical tools: Aristotle's "Rhetorical Triangle" for analysis of verbal and visual rhetoric; the use of search engines and Internet tools such as `whois` and `nslookup` to help establish website ownership and hosting information; and crowdsourcing tools such as Web of Trust as a launching point for delving into the experiences, discoveries, and opinions of other Internet citizens. For example, the effects of methylmercury in food—especially fish—is a contentious and controversial topic. Powerful forces in government, health sciences, and the food and fisheries industries espouse their often-differing viewpoints with vigor using a variety of media, particularly the web. How can the average reader make sense of the clash of studies, claims, and counterclaims appearing on myriad websites of these disparate groups? In the context of my class I asked: How can I teach students enough about rhetoric to be useful in a compact manner?

### Aristotle's "Rhetorical Triangle": A Simple Framework for Rhetorical Analysis

In today's media environment of cynical public relations and hyperbolic political communication it's more important than ever that consumers of information be armed with mental tools for evaluating communication. The three points of Aristotle's "Rhetorical Triangle"—Pathos, Ethos, and Logos—offer a deceptively simple but applicably useful framework for analyzing the how of a message. While no audience can fully know the mind of an author, considering the way a message is composed can offer important clues about the author's motivation and intentions, as well as providing cues for how trustworthy that message is.

The three points of the Rhetorical Triangle are:

- Pathos: the appeal to emotion
- Ethos: the appeal to authority
- Logos: the appeal to reason

Each can be interpreted in several different ways, the core of which I will attempt to describe briefly:

Pathos, the appeal to emotion, is the most closely linked to the sensibilities of the audience. The classic example is the (sometimes ironic) cry of, "Won't someone think of the children!" It reaches directly toward the fears, hopes, and pride of the audience.

Ethos, the appeal to authority, calls upon the audience's impression of the author or those who support the author. It is an appeal to the good name, expertise, and trustworthiness of the one supplying information or opinion.

Logos, the appeal to reason, is the message, or argument, itself. Rhetoric that depends on logos stands on its own informational merit, regardless of the standing of the speaker or the feelings of the audience.

When discussing in class these three points of Aristotle's rhetoric, the group invariably comes to a consensus that, as far as evaluating for truth on the whole, information conveyed with logos is usually the most reliable, ethos less so, and pathos the least reliable. Perhaps it is only an artifact of our scientific-minded culture, but this seemingly pervasive attitude values the repeatable result of a scientific experiment above the say-so of a proclaimed (or self-proclaimed) expert, and both above emotional claims and hyperbolic headlines such as: "Are Vaccines Killing Our Children?" (Mullins, 2009).

University students I have taught respond dynamically and creatively when using this framework to critique the text and design of websites examined in class. In discussion students grasp immediately that most textual or stylistic choices can be placed into one or more of the three categories. It's also worth noting that the class quickly realizes that any piece of truly effective mass communication needs to rely on more than one of the sides of the Triangle. A text that depends purely on logos—such as a mathematical proof or a scientific paper—comes across as dry, lifeless, and meant solely for experts; a text that only depends on ethos raises the hackles of the independent-minded and skeptical reader; a text comprised of pathos can seem naïve and sometimes appears to invite cynical mockery. For example, here is an excerpt from the article whose headline is mentioned above:

> Are vaccines behind the many illnesses our children have today? That is one theory that is going around. Robert Kennedy, Jr. is one of those that question if our children are in danger of their vaccines.
>
> There have been advocates [A] against vaccines because of the threat of autism. They are generally looked at as crackpots in the media, but could they be right? Robert Kennedy, Jr. points out that in areas such as Amish communities and homeschoolers that shun vaccines the rates of autism are almost non-existent. Why then are those rates

> so high in the vaccinated child groups? Natural News [B] reports: "The greatest lie ever told is that vaccines are safe and effective," said Dr. Len Horowitz.
>
> There is a theory that the dangerous preservatives like mercury containing thimerosal within the vaccine hold a serious risk to our children. Could the massive amount of vaccines that children receive during their childhood be poisoning them? At date most children are given about 22 shots before they leave the fifth grade.
>
> In the United States roughly 97% of the population [C] has had their vaccines by the time they reach school age (Mullins 2009).*

Even in this short passage we can see evidence of all three sides of the rhetorical triangle at work. The passage starts with a reference to sick children, always a tug on pathos (whether intentional or not). Next, weakly, we have ethos in a "theory that is going around"—a wimpy tug on consensus—that the author bolsters with logos (a factoid on the lack of autism in vaccine-shunning Amish communities) that is in turn backed with more ethos (the factoid is cited to Robert Kennedy, Jr.). A report from a website called Natural News is ethotically cited, and that citation is both pathetic (in the technical sense, mind—"the greatest lie ever told" is quite alarmist as a phrase) and ethotic (a doctor is quoted). The last paragraph is focused on logos, beginning with the "theory" (a word loaded with logos) that "dangerous preservatives like mercury containing [sic] thimerosal" poses a threat to American children.

I especially like this last paragraph for the purpose of our analysis. It's generally composed in a logical voice, employing a communication mode readers will automatically associate with logos: theory, number, statistic, percentage. Note, however, how the subtle insertion of pathos makes the thesis *fait accompli* in the mind of the reader: To paraphrase, dangerous substances are a danger to children! The rhetorical question in the middle further raises the reader's alarm since as soon as the reader has absorbed that thimerosal is risky and dangerous, he or she is presented with "massive ... poisoning"—strikingly loaded words out of context, but softened considerably within what could almost be considered an aside, so the reader—forgive me—absorbs a massive dose of pathos that calls no more attention to itself than a tiny pinprick. The kind of insight that makes this type of analysis particularly pleasurable is when I realize that while I cynically pry apart the author's writing searching for a core of fact and deconstruct her methods of phrasing, I cannot help but admire its crafty construction as a piece of persuasive communication.

It's also useful and interesting to examine what an article links to when performing an evaluation. In the excerpt above the author appears to be using hyperlinks to other websites as a form of conceptual citation. At root, citations are ethotic references, lending credence to an argument by dint of someone else having said so. Ethos can be the most tangled leg of the rhetorical triangle in this regard; the puzzle of ethos lies in evaluating whether the authority that's called upon to lend its aura to the argument is as correct and worthy as the author intends. The researcher must

* On the quoted web page, text A links to a page on www.globalresearch.ca, "Centre for Research on Globalization" discussing Vaccine Nation, "a film by Gary Null, PhD"; B links to an article on www.naturalnews.com ("Natural Health, Natural Living, Natural News") titled "Vaccines Exposed: A Hidden Crime Against Our Children"; C links to the "intro" page for www.thedoctorwithin.com, which lists some (uncited) statistics but not the 97% statistic mentioned in the article above.

then learn something about that authority. For example, what gives Robert Kennedy, Jr.—certainly a recognizable name—the credence to question the safety of thimerosal? A quick search reveals the About page on his website uncovers that in addition to belonging to a world-famous family he's also a noted environmental lawyer and author (Kennedy, 2010). The excerpt goes on to cite, via hyperlink, in turn: a website on globalization news and two sites on naturalistic and holistic health. It pays to be curious: to give one example, the article cited from naturalnews.com has a byline describing its author Rami Nagel as a "citizen journalist" (Nagel, 2007). Performing a search for Mr. Nagel's name reveals that he has written two books on health, *Cure Tooth Decay* and *Healing Our Children* (Cheeseslave, 2010).

This kind of investigation can go on and on—part of why investigating ethos can be so difficult. Unlike logos, which should stand or fall on its own, and pathos, which simply calls on what feels right to the reader, with ethos there's a need to check out the sources, and the sources' sources, possibly leading to an exponentially branching search akin to a genealogist trying to track down all several dozen million descendants of King Edward III. On the other hand, a diligent investigation just a few levels deep usually reveals enough to give a feel for what kind of laurels a claim to ethos is resting on—ranging from peer-reviewed scientific papers to "ethotic loops" of bloggers and self-proclaimed "experts" all citing each other.

## Internet Sleuthing

Besides noticing what is being said and how it is presented, it is also useful to find out more of the who, where, and when of a website. We must use technical instead of rhetorical tools to investigate these questions. The main tools I use to find out more about a website—namely where it is hosted, who owns its domain name (and where that name is registered), and how long ago the page was modified—are `whois`, `nslookup`, Google (though any search engine will do), along with the web browser itself.*

Websites are, for all intents and purposes, just files on a computer somewhere. The point of using these tools is to try and find out who is responsible for those files, where the computer is located, and what companies or individuals are responsible for them. The Internet is designed to handle requests transparently—so it doesn't functionally matter to a user if a site is located in Hong Kong or Vancouver—but the ability to look under the hood a bit can be useful when trying to evaluate information.

Let's start with the most familiar aspect of Internet sites: their domain names. A domain name is what a person is used to typing into a web browser (e.g, "facebook.com" or "www.wm.edu"). A common example would be something simple like "www.google.com." In this case the suffix (".com"—which can also be referred to as a "top level domain" or "TLD") tells us that the entity is (ostensibly) commercial

* Speaking of search engines, as an aside, it's worth mentioning that searching for mercury—of the elemental and hazardous variety—can be hard. Cars, newspapers, a rock star, a deity, and a planet all share the mercury name. Finding just the right information on the right website can be a challenging exercise all by itself.

in nature. Its domain name is "google" and the host we are trying to reach is called "www": the most common host name on the web. You can think of this as "a system called 'www' belonging to a company called 'google.'" The entirety of www.google.com is called a "fully qualified domain name," or FQDN, because it refers to a specific system on the Internet. (Web addresses are often prefaced with the "http://" construction. HTTP stands for "HyperText Transport Protocol." In short, this prefix simply tells the web browser that the target is a web page.)

Here's a slightly more complex example: I just searched Google for "mercury hazards" and the top search entry was a web page entitled "Mercury Health Hazards" hosted at orf.od.nih.gov. A short bit of sleuthing and thinking about the site address yields fruitful information. The domain "nih.gov" is owned by the National Institutes of Health, a branch of the US government (its governmental status is revealed by the ".gov" suffix). The "od" portion (i.e., od.nih.gov, apparently "Operations and Development" according to the header graphic on the site) is what's called a "subdomain," a subdivision of the organization's domain, which is a convention often used when different parts of a domain's naming space are broken down into separate regions of control, like subdivisions of a corporation. The initial portion, "orf," is the hostname, the name designated to a specific computer system. (ORF stands for "Office of Research Facilities," again gleaned from the header graphic of the website.) Our research has revealed a little already—since most people wouldn't know that the top web hit for "mercury hazards" on Google is a site run by the Office of Research Facilities, a branch of the Operations and Development division of the National Institutes of Health!

As mentioned, there are a number of different domain suffixes, or extensions. Ostensibly, each one carries some connotation, though most of them can be registered by anyone. For example, the original domain extensions were .com (for commercial enterprises), .net (for networks and network providers), .org (for organizations), .edu (for colleges and schools), and .gov (for US government and its institutions), along with a two-letter extension for each country (.ca for Canada, .us for the United States, .cn for China, etc.). Some countries are proprietary with their domain extensions, while others sell domain names within their country designations to make revenue; for example, the Christmas Islands (.cx) and Tuvalu (whose .tv extension is popular with television stations). In practice, though, anyone could register a .com, .net, or .org domain at their pleasure. Other TLDs, however, such as .edu and .gov domains are protected from being registerable by just anyone—for example, only accredited schools can obtain .edu and only government entities can obtain .gov addresses. Complete lists of domain extensions, their meanings, and relative restrictiveness can be found easily with a search engine. Really, though, domain names and web addresses are only text-based shorthand to make navigating and managing Internet sites easier for people. Just as it's easier for most people to remember someone's name than it is to remember their phone number, domain names serve largely as a mnemonic for people. The computers they refer to are actually found by other computers via their IP (Internet Protocol) addresses. IP addresses are like phone numbers, complete with area codes and exchanges, so that computers can route to each other and thereby communicate.

One quick note on the tools: when I use the tools mentioned in this chapter (`whois`, `nslookup`, `host`, etc.) I'm using the traditional command-line versions on a UNIX system. (Their short, mashed-together names are a reflection of their origin as typed commands.) Macintosh users can also access these command-line tools by launching the Terminal app since Mac OS X is based on a UNIX variant. Windows users have a command-line version of `nslookup` but not `whois`. However, anyone using any operating system can access web versions of all these tools from various sources such as DNSstuff.com, domaintools.com, and lots of others. Additionally, most (all?) registrars offer an online `whois` tool—if nothing else so that a person can check to see if a domain is available.

### Using `whois` to Find Out Who Owns a Domain

As I mentioned above the most common use of `whois` is to check and see if a domain name is available. The `whois` tool provides a lot more information than that—including who is responsible for that domain, along with contact information; how long the domain has been around; how long it's been since the information on the name has been updated; how long until its registration expires; and what name servers on the Internet are responsible for the names of the hosts that reside within the domain. Contact addresses for various roles are provided, usually including a billing contact (or registrant), an administrative contact, a technical contact (for Internet-related issues originating from the domain), and often these days an abuse contact (for reports of spam, electronic break-in attempts, and other Internet malfeasance).

For a running example I'm going to do a short investigation of the site www.arin.net.* Checking the `whois` information is an easy way to get information on who owns the domain corresponding to the page. Note, we're looking for the `whois` information for arin.net, not www.arin.net, since the former is the domain (to which `whois` data applies) and the latter is a hostname (the text address of some machine on the Internet). It's also worth mentioning that many websites will respond without prefixing "www"—but that's a contrivance to make things easier on the user. You may notice that if you type in "google.com" in your browser it gets corrected to http://www.google.com/ when the page loads.

The results in Figure 7.1 answer several questions a researcher might have about who's behind a website. We learn here, if we haven't already learned from the site itself, what ARIN stands for along with the street address of its headquarters and some contact numbers. We can also see that its registration expires (usually, how far ahead the registrant is paid up) in 2019 and has been around in this particular instance since 1996. The "Database last updated" line tells us how recent our information is. Finally, we are provided with a list of domain servers; these are computers on the Internet that are responsible for providing all the information for arin.net to

* I was going to provide an example of a mercury-related site, but unfortunately republishing `whois` data is forbidden by ARIN's Terms of Use, and anyway is subject to change. ARIN, the American Registry for Internet Numbers (which owns the `whois` data for the major US TLDs) graciously agreed to allow publication of its own `whois` information for the purpose of this chapter.

```
Registrant:
American Registry for Internet Numbers
   3635 Concord Parkway, Suite 200
   Chantilly, VA 20151
   US
   Domain Name: ARIN.NET
   Administrative Contact, Technical Contact:
      Ryanczak, Matt            noc@ARIN.NET
      American Registry for Internet Numbers
      3635 Concord Parkway, Suite 200
      Chantilly, VA 20151
      US
      703-227-9869 fax: 703-227-0671
   Record expires on 20-Dec-2019.
   Record created on 19-Dec-1996.
   Database last updated on 16-Aug-2010 17:48:49 EDT.
   Domain servers in listed order:
   ALOE.ARIN.NET                 192.12.94.32
   C3.NSTLD.COM                  192.26.92.32
   D3.NSTLD.COM                  192.31.80.32
   DILL.ARIN.NET                 192.35.51.32
   NS1.ARIN.NET                  199.212.0.108
   NS2.ARIN.NET                  199.71.0.108
   M3.NSTLD.COM                  192.55.83.32
```

**FIGURE 7.1** The result of "`whois` arin.net" (note: some boilerplate text has been removed from the result, leaving only the meat of the data).

the Internet at large—what the numerical addresses of its hosts are, including the ever-important www, where its e-mail should be delivered, etc. In this case there are a lot of them, providing a lot of redundancy in case one or more go down or become otherwise unavailable. Most domains only have two or three.

## Using `nslookup` and `whois` to Find Information about Hosts

`Whois` doesn't just supply information on domain names; the same system provides information for every Internet network in the world. As I mentioned above it is possible that a company may have a name registered in the United States, but that has no bearing on where the server that hosts its website is located. To check we can use `nslookup` to find the numerical address of their web server, then use `whois` again to find out about what network it's on. If I run "`nslookup` www.arin.net" (this time we want the host, not just the domain), I get the result shown in Figure 7.2.

The "Server" and "Address" lines don't have any bearing on www.arin.net; they indicate where the `nslookup` command queried for the information is. In this case it's the server I'm on; the IP address 127.0.0.1, called the "loopback address," is a special IP address that always means, as far as a computer is concerned, "me." The "Non-authoritative answer" line is also an indication about the server queried. It means, "I am not authoritative for this information; I got it from elsewhere." If I

```
Server:         127.0.0.1
Address:        127.0.0.1#53
Non-authoritative answer:
Name:   www.arin.net
Address: 192.149.252.76
Name:   www.arin.net
Address: 192.149.252.75
```

**FIGURE 7.2** Results after runing "`nslookup` www.arin.net."

had directly queried one of the nameservers listed in arin.net's `whois` data, this line would be elided.

The result we get is somewhat interesting in that we got two separate answers. The hostname www.arin.net has not one, but two different numerical addresses. This is only more redundancy; ARIN has two separate addresses serving its website, and since some computers may hit one and some hit the other it provides a bit of traffic management as well. Fortunately, both addresses are part of the same network (only the final number differs in this case) so we don't end up with more networks to chase down. To find out information about where these hosts are, we'll run the first address against `whois`. Running "`whois` 192.149.252.76" gives us the results seen in Figure 7.3.

One telling part of all this information is that the network is owned by the same organization that owns the domain. That's an unusual occurrence; most of the time a network is owned by an Internet Service Provider (ISP) or a company that sells web space. Any organization that has registered its own address range is either itself an ISP, a hosting provider, or a company, government, organization, or school large enough to have a serious need for an Internet presence. Another useful result is the contact information for the network's responsible parties. Again, usually this would

```
NetRange:       192.149.252.0 - 192.149.252.255
CIDR:           192.149.252.0/24
OriginAS:       AS10745
NetName:        ARIN-NET
NetHandle:      NET-192-149-252-0-1
Parent:         NET-192-0-0-0-0
NetType:        Direct Assignment
NameServer:     SEC1.AUTHDNS.RIPE.NET
NameServer:     NS1.ARIN.NET
NameServer:     NS2.LACNIC.NET
NameServer:     SEC1.APNIC.NET
NameServer:     NS2.ARIN.NET
RegDate:        1997-11-05
Updated:        2010-08-20
Ref:            http://whois.arin.net/rest/net/NET-192-149-252-0-1
OrgName:        American Registry for Internet Numbers
OrgId:          ARIN
Address:        3635 Concorde Parkway
Address:        Suite 200
City:           Chantilly
StateProv:      VA
PostalCode:     20151
Country:        US
RegDate:        1997-12-22
Updated:        2010-01-09
Comment:        For abuse issues please see URL:
Comment:        http://www.arin.net/abuse.html
Ref:            http://whois.arin.net/rest/org/ARIN
OrgTechHandle: ARIN-HOSTMASTER
OrgTechName:   Registration Services Department
OrgTechPhone:  +1-703-227-0660
OrgTechEmail:  hostmaster@arin.net
OrgTechRef:    http://whois.arin.net/rest/poc/ARIN-HOSTMASTER
OrgNOCHandle: ARINN-ARIN
OrgNOCName:   ARIN NOC
OrgNOCPhone:  +1-703-227-9840
OrgNOCEmail:  noc@arin.net
OrgNOCRef:    http://whois.arin.net/rest/poc/ARINN-ARIN
```

**FIGURE 7.3** Results after running "`whois` 192.149.252.76"

be a different organization than the owner of the domain name, and instead reveals the people responsible for the network the website's server resides in.

As far as information evaluation goes this kind of result is the crème de la crème of legitimacy, at least as far as authorship goes. They are certainly who they say they are. This type of investigation can be very useful, though, when you're not sure or just want to know what kind of hosting they are using. For example, let's say an e-mail message you receive, claiming you are due some money, points you to a page claiming to represent the government of Nigeria. You don't even need to open the page (what if it's a scam and the page is full of viruses?). Run `whois` on the domain to see who registered it, `nslookup` the hostname, and then run `whois` on the result to see where it's hosted. If our hypothetical site were registered by an American but resides in Russia you can at least be sure it has nothing to do with the Nigerian government!

Here's an interesting example from the realm of mercury-related websites. In a search for "eating fish mercury" (I find that when doing Internet searches on Google, a keyword-only approach often works well) one of the results I find is a site called "FishScam" at www.fishscam.com. The site is slick and professional in appearance. Its front page advertises a book called *Tuna Meltdown*, and the blurb for the book talks about how impoverished children are "victims of public-health malpractice by green groups and the federal government, which have been issuing dire but exaggerated warnings about harmless levels of mercury that have always been present in ocean fish" (fishscam.com, 2010). The site also has news articles with titles such as "Activist Fish Report Full of Chum" and "Experts: Eat Your Seafood (Without a Side of Scaremongering)." The site is expounding a quite different narrative than most other sites (and publications) on the same topic. Of course there are always two sides to every story, and the critical researcher will want to know: whose side this is?

Checking `whois` for the domain (as of this writing in August 2010) reveals a curious result: we can see that the domain is registered at GoDaddy.com (one of the largest commercial domain registrars and web hosting providers), but the contact information provided indicates "Private Registration" by a site called "domainsbyproxy.com." According to its home page DomainsByProxy provides a privacy service for domain registrants. The site asks the potential client, "Did you know that for each domain name you register, anyone—anywhere, anytime—can find out your name, home address, phone number and email address?" Since the law requires domain registrants to provide valid contact information, DomainsByProxy acts as a forwarding service, analogous to postal mail forwarding services for those who wish to correspond by mail in relative anonymity.

So where do we go from here in our research? Checking `nslookup/whois` for its IP indicates a dedicated hosting provider in Atlanta, Georgia (again, note this information is subject to change by the time this essay is published), but that doesn't really tell us a lot, other than fishscam.com isn't running its site on servers in its own basement. But it's important not to miss the forest for the trees: on the fishscam site itself its "About" link (at the bottom of the page) indicates that "MercuryFacts.org is a project of the Center for Consumer Freedom, a nonprofit coalition dedicated to promoting personal responsibility and protecting consumer choices." But we are on fishscam.com, not mercuryfacts.org. Go there and you'll see a nearly identical site using the mercuryfacts.org name. The domain is also behind the "Domains by Proxy" identity wall. With `nslookup` we see that this site is on the same server as fishscam.com. When I check consumerfreedom.com, however, I do get real contact information in the `whois` data, and this site is also on the same server. Furthermore, the consumerfreedom.com site has a whole list of links to other sites, including:

- ActivistCash.com
- AnimalScam.com
- CSPIScam.com
- HowMuchFish.com
- MercuryFacts.org
- ObesityMyths.com
- PhysicianScam.com
- SweetScam.com (Center for Consumer Freedom, 2010)

That's a lot of scams! Browsing these sites reveals a common theme—in short, that devious environmental groups and physicians are trying to scare consumers into not buying and eating whatever they please. What we've found at this point, I believe, is a public-relations campaign.

### Putting It All Together

So what's a reasonable evaluation of fishscam.com? We've used technical tools and web searching and browsing to link FishScam to an entire bevy of websites that all originate from the same organization. Browsing them we find a common rhetoric, too, appealing to pathos with heavy-laden words such as "scam," "scaremongering," "myth," "spin," and "hype," while also making a claim to ethos through concepts such as "facts," "reports," "scientists," and "studies." The careful reader might also note that the negative terms are always applied to the opposing viewpoint and those who espouse it (that mercury in seafood is hazardous enough for concern, for example), while the ethotic terms are always applied to items supporting the message the sites want to provide. There is logos, of course, but most if not all of the logic depends on the reader investing belief in the authority of the sources and facts cited. For me, and usually for my students, the real warning bell is the bombastic use of loaded words. Overreliance on pathos is very often a mask for a weak argument. Fishscam.com makes as many negative claims about their ideological opponents as they make positive claims about their message—not quite to the level of a negative political campaign, but certainly notable.

## OUTPUT: COMMUNICATING MERCURY IN THE DIGITAL AGE

Imagine a luminous silver drop spreading until it is broken into myriad tiny fractions, dispersing until each fragment is spread far and wide across the globe, where hundreds or thousands or millions of people absorb it one minuscule portion at a time. Some is lost after it is absorbed, but some remains inside—and as each individual absorbs more it builds to become a more prominent part of its host. You are probably thinking about mercury, but here I am talking about information.

The difference, of course—the broken part of the analogy—is that information is infinitely reproducible, while a droplet of quicksilver is finite; both are "safest" when contained in quantity in a proper receptacle. Diffused throughout an environment, each becomes very difficult to gather together again, lending mercury its danger and information its power to move from just being a fact or a collection of facts into a part of zeitgeist and culture. This section will explore some of the ways information about mercury makes that transition, moving from the safe containers of compendia, government reports, and industry studies and into the minds of the millions of global citizens who are affected by mercury in our world. My goal is to encourage those on the forefront of mercury studies—or indeed, any kind of useful study—to put diffusion, the dissemination of information, to work using digital media of all kinds. The focus is upon small packets of information, of the kinds that can grab attention and be absorbed easily and readily—for as enough tidbits of knowledge are digested by information consumers, curiosity becomes interest, and trivia begin to form a body of knowledge.

## CONTEXT: A BRIEF HISTORY OF DIGITAL MEDIA

### Pull

In the early days of the web, like the very early days of publishing, there was not much material to choose from. The web was so small the Mosaic browser had a link to a periodically-updated "What's New" page built into its menus. The new web was something of a curiosity, and users would trawl thc Internet for interesting, useful, or entertaining sites. Sites became popular through word of mouth (or at least the email equivalent), and nearly every page had a list of links to sites that appealed to the author. As the web grew—and search technology along with it—web readers had to develop a new set of skills to find relevant sites. Where today an online mercury researcher might share interesting finds with like-minded readers in real time on a dynamic social bookmarking system such as Digg or reddit, or perhaps a purely social platform such as Facebook or Twitter, in the early days of the web he or she would create a static web page featuring a collection of mercury links. Web rings, collections of sites linked via a header or footer managed from a central server that would link to each other in turn (thus, the "ring") was as dynamic as it got. (Incidentally, I searched for "mercury Web ring" and found several Queen- and car-related rings, but none for elemental or environmental mercury.) Presuming the information was available out there the reader, like anyone wandering the stacks in a library, had to go and find it. And like a hand reaching for a book on a shelf, this type of information retrieval is termed "pull."

Pull techniques are as old as information collation itself. The problem with them is that the seeker has to locate what is sought and then retrieve it. A twelfth-century monk seeking a particular scroll would need to locate it within a series of rooms, shelves within that room, and in a particular stack on a particular shelf. Likewise a user of a scientific database may seek a particular series of data, but in this case she or he must formulate the proper query to request it—sometimes in an arcane language. The point is that pull techniques are rather difficult, and even when they are relatively facile they still require some effort on the part of the seeker, right down to the formation of the idea that he or she is looking for something in the first place. What about good, pertinent information that a seeker didn't know to look for? It is lost in the sea of information, unknown.

As the web blossomed into millions of pages search technology became the most important arbiter of information. Early search engine technology depended upon keywords and, to some extent, crunching the content of web pages. However, as individuals and companies began to make money from websites, some chose to game the system by inserting decoy content into web pages: a search for "environmental mercury" could lead just as easily to a salacious advertisement as it could to an Environmental Protection Agency publication. To avoid such pedestrian subterfuge, search engines (beginning with Google's PageRank system) began using the metric of how often pages link to each other as a key for sorting sites by relevance. Certain widely-referenced sites could be easily recognized as nexuses of information on their topics. Suddenly, popularity became more than notoriety in certain circles; it became logarithmic. In this way a really useful collection of links on environmental mercury could rise to the top of the search heap, by dint of the ranks of the sites it links to, how many of them linked back to the list, and how many other sites would link to that resource. Like the economics of capital—where the vast majority of money drifts to a small percentage of hands—most seekers of information are therefore referred again and again to a relatively small number of sites compared to the countless millions of individual web pages.

## Push

People, however, are individual. They are not just wandering, sheepish consumers who will always be satisfied with the top-ranked entry from a web search. Sometimes something very, very specific is required—whether a topic, or entertainment with a particular style of humor, or the ruminations of someone like-minded. Recognizing that pull techniques are effortful, content seekers needed a way to not only return to a discovered resource (bookmarks have been around since the very earliest web browsers) but to also stay in touch with those resources and be notified when new, relevant information was available. Push technology was born as a response to this need.

The most widely-used incarnation of push techniques for websites today comes in the form of "feeds" of one type or another, with names such as RSS ("Really Simple Syndication") or Atom. You may have noticed the familiar icon on websites you've perused. This icon represents a common way of indicating that there is a feed available for that site. The feed is really just a link to a specially formatted web document—one that's not meant to be read by a person but checked by a program called an aggregator to see if there's anything new. When the feed is updated the user's feed reader software displays a pointer to the new content, usually in bold or with some other formatting to indicate that that content is unread. In these cases, content—stories, comics, or photographs—is chunked into units on a single page, with the newest content at the top; older content gets pushed downward. This format of online publishing is referred to as blog format ("blog" is a contraction of "weblog"). There are dozens of aggregators available for audiences to use to manage a collection of feeds, both on the web (two popular web applications are Netvibes and Google Reader) and with standalone software that runs right on the user's computer (such as RssReader).

When a feed provides links to sound files (either spoken word or music) it is called a podcast, or in the case of videos, a video podcast ("podcast" is a pun on "broadcast," incorporating the name of the popular Apple iPod™ line of video and music players). Radio and TV shows are commonly available as podcasts, though the popularity of the technology has led to media outlets providing many hours of content that is solely online. Apple's iTunes application is probably the most popular way to subscribe, manage, and listen to audio and video podcasts—whether or not the user has an iPhone or iPod.

Since much of the web has moved from static pages to dynamic content platforms and practically every such platform supports RSS, you can find plenty of feeds to subscribe to for mercury-related information—ranging from environmental science magazines and journals to websites and blogs on topics ranging from alternative medicine to social activism. You can also use a service such as Netvibes to generate an RSS feed of a web search to be notified when the search becomes aware of new or more highly-ranked content.

## Publish

The history of the web, like any phenomenon that captures global attention and imagination, is a parable of populism. Originally the web was conceived as a primarily text-based medium to be used by researchers for the association of different chunks of information. If you wanted to publish on the web in the early 1990s you'd have to learn both its native tagging system, HTML (Hypertext Markup Language), and the arcana of managing files on the servers that hosted web content—if not the servers themselves! As the web gained in popularity, web authoring software became available, first providing markup and then graphic design. As Internet speeds grew, sound and video became both more commonplace and higher in quality. As more people wanted to create websites, the market for self-contained, all-inclusive web publishing software grew. Of course as web design became more complicated, more complicated tools to design professional-quality sites from scratch also became available. But what about those who just wanted to publish their thoughts, writings, art, music, or video without having to go through the trouble of becoming web designers?

The desire to blog (in the verb form), to publish text, sound, and video, led to the development of content management systems (CMSs) that make it quite simple to publish online. A CMS is nothing but a kind of web authoring software that lives on somebody's web server instead of the author's computer, meaning that, to publish, all a potential author needs is a web browser (every computer comes with one) and Internet access. Instead of memorizing scads of HTML markup tags, posting to a blog is no more arduous than writing in a word processor. Sites such as LiveJournal and Blogger provide a blogging outlet that is easy to use, visually customizable through the use of prepackaged themes, and, best of all, free of charge (albeit subsidized by advertising, as are most websites that provide a free service). WordPress is another popular open source program that started out as blogging software but grew into a fully featured CMS capable of managing sites of all kinds; it can either be run on an author's own web server or accessed freely on wordpress.com.

In addition to the simplicity and automation provided by blogging software most offer an additional tagging feature that makes it easy for authors to categorize their posts and for readers to find them using keywords. A person can find many environmental blogs on the web whose contributors sometimes cover mercury-related issues, but if you subscribe to several of them it could become arduous to scan them all periodically for mercury mentions. Tagging alleviates this problem by creating a mercury-focused version—assuming the bloggers tag their articles in a useful way. Two good examples I found come from journalist blogs: one from *The New York Times* (<http://green.blogs.nytimes.com/tag/mercury/>) and one from another paper in New York, the *Times Union* (<http://blog.timesunion.com/green/category/mercury/>). Both sport environmental blogs with a useful tagging system. While I didn't find an explicit link to a category-based RSS feed on those pages, I was able to guess correctly how their software (WordPress or at least WordPress-like) manages addresses and generate one by tacking "/feed" on the end of the URL (e.g., <http://blog.timesunion.com/green/category/mercury/feed>). By asking my RSS reader to track that address I now have a single place to check whether new mercury-related articles have appeared on those blogs.

## Today's User-Driven Social Web

### User-Driven Content

Once blogging becomes a business an important consideration is how to increase revenue and minimize expense. Remember that the free parts of the web are usually advertisement-driven, in which case traffic equals eyeballs, and eyeballs on advertisements (or, if you're small-time, actual clicks on advertisements) equals revenue. It's a numbers game—some small average percentage of users will click on advertisements so the way to increase clicks is to increase traffic. The margin for all but the most popular sites is thin (web ad revenue is measured in cents per click) so paying a freelance writer for content is often out of the question. The solution? Have the users provide the content! Their recompense is but notoriety and the pleasure anyone feels upon being listened to.

Advertising is how blogging sites such as Blogger and WordPress can offer a free service for individual publishers. It is also how many topical sites sustain themselves. One early example of such a topical site is called Slashdot, a community-driven site in a blog format that focuses on open source software, Internet technology and issues, science, technology business, and the like. (Its motto is "News for nerds. Stuff that matters.") Most of the content on Slashdot is in the form of links to articles elsewhere on the web along with a summary. Each entry also features a threaded comment page where other readers—anonymously or not—can comment on it. "Stories" (as they are called) are culled by editors for selection on the front page. But here's the real brilliance of the model: most of the links to interesting items are contributed by the readers! This successful format has become widely adopted by many sites, serious and frivolous, technical and entertaining. Sites with thousands—or millions—of participants benefit from the users' diverse browsing habits and penchant for discussing topics of interest, prospering by the attention of a big population of free-ranging editors-at-large at no additional cost to the site.

Although I couldn't find an environmentally focused, community-driven mercury website, such a site would not only be a great clearinghouse for information but would provide a community for like-minded people sharing and discussing a common interest. Of course more general social sites such as Slashdot (or Digg, reddit, etc.) support a robust enough tagging system that a researcher or casual browser can focus more specifically on mercury or the environment.

### Net of Society/Society of Net

The confluence of the popular growth of the Internet and the democratization of online publishing has led directly to complex webs of interaction between people online. These threads of interaction cross technologies as easily as they do boundaries of countries and social class. Another form of website (also usually a business) focusing solely on this social function gained popularity in the mid-2000s—the advent of social networking. Social networking sites, such as Myspace, Facebook, and LinkedIn, differ in look and feel and, somewhat, by audience, but commonalities include the notion of "friends," people with whom a user is connected in some way and with whom that user can share information in a sphere between completely public and completely private; modes of public, semi-public (i.e., "friends-only"), and private communication between users; a common platform for sharing media such as embedded audio and video, and web links; and a form-driven profile where users can share information about themselves. Sometimes users inadvertently share too much information; in fact, in the year 2000 hitting "reply all" to an e-mail was the stereotypical method of sending private and potentially embarrassing information to the wrong recipients—in 2010 the stereotypical foible is misunderstanding the access permissions for an item on a social networking site and often a compromising photograph gets shared among friends, and among their friends, and among their friends, etc. Likewise, all social networking sites provide some way to see when one's "friends" have posted something—a status update, blog post, link, or embedded video. As noted previously, friends can share anything you have shared with their friends, leading to so-called "memes" (a word far looser in its popular meaning than in its academic sense), spreading like tiny wildfire fads that travel orders of magnitude faster than any hemline or tie width could ever hope to oscillate.

Other kinds of social sites focus on specific types of media. Photo-sharing sites such as Picasa and Flickr allow photographers and graphic designers to share images; social bookmarking sites such as Digg, Metafilter, and reddit allow users to share and discuss web links—along with automated data mining to allow anyone to see what sites and topics are at the type of the zeitgeist (remember, popularity is logarithmic). YouTube is by far the most popular site for sharing video. LibraryThing combines social networking for bookworms with a surprisingly feature-rich library cataloging system. There's even a social networking site called Etsy that provides networking opportunities and a sales platform for people who do home crafts of every imaginable variety. And there are hundreds—thousands?—of other social networking sites, some popular and some small, depending on the specificity and rarity of the niche each is trying to fill, each featuring a community with its own type of flavor.

Twitter is another social-networking service that's in a class all its own. It is a kind of confluence of the SMS text-messaging used between mobile phone users and the web. Another way of looking at it is social networking without the website—one tiny bit at a time. Twitter users can publish textual messages of 140 characters or less often including web links shortened by hyperlink-shortening services such as bit.ly, one at a time. These messages, called "tweets" (verb form "tweeting"; users sometimes even refer to themselves as "twits"!), can be seen by anyone who browses to your Twitter page on the web, but that's not quite the point. Instead of "friends" (everything you post to Twitter is public, save direct messages to other users) Twitter employs the metaphor of "followers," and everyone who "follows" you gets your tweets pushed to them via their Twitter page, but also possibly to a dedicated Twitter desktop widget on their computer or their smartphone or even, via syndication, to the same aggregator page or software they use to keep up with blogs and other web resources. Tweets can also be "hashtagged" by prefixing one or more key words with a hash sign (e.g., #mercury) so anyone searching for that topic can see your tweet along with all others containing that tag.* Likewise you can draw the attention of any Twitter user by prefixing their username with the "at" symbol (e.g., "@ billgates—pls help us fund the 2010 Int'l Mercury Expo")—that user will then see your tweet in their feed.

### Interconnected Social Media: Stirring the Pot

The present-day apotheosis of social media is reached not just by the way users interact with the various services but instead in the ways these services interact with each other. Tweets can be pushed to one's Facebook status, and blog posts can be announced via Twitter. Photos on Flickr can be shared on every service you subscribe to. Add a video on YouTube to your favorites on that site and the occasion can be tweeted, displayed on Facebook, mentioned in your blog, and sent to an RSS feed without any extra effort on the user's part if everything's configured to do that. Any of these media can be sent or received via smart mobile phones such as iPhone and Android. With so many interconnected people using so many interconnected services tidbits of information can spread like, well, not wildfire, for social networking has outstripped that metaphor for fast dissemination. Perhaps we need a new metaphor.

Some trends are clearly identifiable considering the spread of information on the Internet. Information chunking has morphed from large scale (sites, pages) to small scale (blog posts) to micro scale (tweets and status updates on social networking sites such as Facebook). The web has also increasingly moved from linking information to information to linking people and organizations with each other—while at the same time information has become easier to share utilizing prebuilt web-based platforms designed for ease of use and a high degree of interconnectedness to other platforms, thanks to open standards such as RSS and the sharing of Application Program Interface (API) code between platforms and platform developers. As a result, users of these services are sharing lots of personal information as well as keeping contact with networks of friends and acquaintances—along with famous people, organizations, and companies that capture their interest.

## Inserting Mercury Awareness into the Digital Ecosystem

Advocacy and awareness are hugely important parts of the problem of mercury. I've wondered to myself: for every automobile with an "awareness ribbon" magnet on the back, how many of the drivers have ever considered the effects of industrial and environmental mercury on humans and our ecosystem? As our cultural attention drifts from television and newspapers to digital media, and our cultural attention span shortens and becomes increasingly a battleground for the competing ideas and sources of information, the researchers, advocates, environmentalists, and health-care providers who generate and provide information must make efforts to reach the new audiences. Just as mercury quarantined in a vial does no harm, information

* This is another example of how searching for mercury can be challenging. Twitter supports an advanced feature where you can modify a search for text or a hashtag, including what you don't want to search for using the minus sign. I filled the maximum length of the query with things like "-Freddy," "-Queen," "-San Jose," "-Cougar," etc., and still got many tweets I wasn't interested in!

cloistered does no good. As individual awareness becomes societal awareness—passing from monolithic research into a collection of information shared by the society of individuals just as information in the digital age has metamorphosed from monolithic, static, and rare to more fragmented, fluid, and shared—a shift of cultural attitudes cannot help but follow.

In conclusion, and to return to my purpose in writing this essay: to become a better information transmitter, become a better information consumer. Being able to critique information and its presentation, knowing where information comes from and its provenance, and being able to judge the impact of that knowledge on how the information should be received means being able to make good choices on how to present and cite information. Knowing the advantages and disadvantages of different media, how they're consumed, and who consumes them means being able to choose which ones are appropriate for which kinds of information. Receiving, judging, and passing on information about environmental mercury—or any other topic—makes a person a node in the web of the exchange of ideas. Go out and explore the digital ecosystem. Teach others to become savvy web readers. And most of all don't just learn—participate. Find—or better, start!—a blog or a Facebook group or a Twitter account or a website or a YouTube channel that interests you. Present your ideas in the most credible way you can. Point out useful communities and information sources to others. And always remember that knowledge—truth, if I may be so boldly old-fashioned—like mercury spreads easily, builds up, and is terrifically hard to remove once absorbed.

## REFERENCES

CheeseSlave. 2010. *How to Cure Tooth Decay with Rami Nagel*. Accessed January 27, 2010, at http://www.cheeseslave.com/2010/01/27/how-to-cure-tooth-decay-with-rami-nagel/.

Kennedy, Robert F. Jr. About. http://www.robertfkennedyjr.com/about.html.

Mullins, K.J. 2009. Are vaccines killing our children? *Digital Journal*. Accessed October 12, 2010, http://www.digitaljournal.com/article/280412.

Nagel, Rami. 2007. Vaccines Exposed: A Hidden Crime Against Our Children. *Natural News*. Accessed December 18, 2007. http://www.naturalnews.com/022400.html.

# 8 Making Mercury Visible
## *The Minamata Documentaries of Tsuchimoto Noriaki*

*Justin Jesty**

## CONTENTS



> The roots of Minamata disease are different from those of other human diseases. ... It is a chemical poisoning, whose destructiveness is at heart a reaction against human beings, caused by [the drive towards] "civilization." ... Since the discovery of the disease, industry has interfered with the discovery of its cause, in a cover-up going all the way to the prefectural and national governments. That this obstruction has fundamentally not been overcome even today, testifies to the fact that Minamata disease is, in addition to being a disease of the human body, a thoroughly social disease.
>
> —**Tsuchimoto Noriaki,** from the production notes to the *Minamata Disease* trilogy, 1973

## INTRODUCTION

Today, Minamata might be well known as one of the most disastrous cases of mercury pollution in history. This outcome, however, was far from obvious to the fishermen around Minamata who first began to suffer the acute effects of mercury in their food in the mid-1950s. Although research quickly began to point toward the effluent from the local Chisso chemical plant as the cause—all the links finally established

* The author would like to thank the Council on East Asian Studies at Yale University and the Yale University Library for making this research possible, as well as Timothy S. George, Abé Mark Nornes, Ann Sherif, and Sharon Zuber for their generous input during the writing process.

in scientific papers by 1961—this knowledge had little noticeable effect. The victims remained in obscurity for years after, their suffering something that few people wanted to hear during Japan's heady high growth of the 1960s. The victims referred to this decade as the "decade of silence." In addition it was a decade of ignorance. The causes of Minamata Disease were successfully obfuscated by a host of counter theories from industry and government scientists, so that few in Minamata, and far fewer nationwide, knew or cared about the dangers of mercury in their environment. Until 1968 Chisso continued to dump untreated organic mercury into the Shiranui Sea and people continued to fish from it. As Timothy S. George has argued, Minamata is not just a story of the environmental and human costs of pollution, but of a corporation bent on hiding its guilt, a government happy to collude, a media in confusion, and "powerful pressures against speaking out and taking action" (George, 2001, 8–9). It is not only a story of devastation but of how public understanding must be painstakingly built and fought for.

Building such an understanding is the work of culture. Through the decade of silence a handful of writers, doctors, scientists, photographers, and filmmakers undertook the work of bringing the story of Minamata and its victims to a wider public, and in the late 1960s they were joined by a growing network of allies from around Japan and the world. Documentary filmmaker Tsuchimoto Noriaki was among these. Between 1965 and 2004 he made seventeen television and film documentaries about Minamata.* This chapter will focus on five films completed in the 1970s that have become the defining feature of his work and have been used around the world as a resource for understanding mercury pollution. Together the films represent one of the most detailed and long-running studies of an environmental disaster on film, and bear witness to the evolving struggle of the victims to cope with the toxic effects of mercury.

The films are still relevant today, first because they document and demonstrate the continuing importance of local knowledge. Knowledge constituted on the margins is often treated as illegitimate: as we will see in the case of Minamata it was a combination of existing prejudice against a marginalized population, questionably constituted official knowledge, and a media committed to balance rather than truth that succeeded in disenfranchising the voices of victims and local researchers. Tsuchimoto once wrote that he and his crew "set out for Minamata in order to get the reverse view, to look out from the small hamlets, at the city, at the prefecture, at

* In chronological order, these are, *Minamata no ko wa ikite iru* (The Children of Minamata Live On) (1965); *Minamata: The Victims and Their World* (1971); *Minamata Report 1:The Central Pollution Board* (1973); *Minamata Revolt: A People's Quest for Life* (1973); *Minamata Disease—A Trilogy Part 1: Progress of Research (1974); Minamata Disease—A Trilogy Part 2: Pathology and Symptoms* (1974); *Minamata Disease—A Trilogy Part 3: Clinical Field Studies* (1975); *Shiranui Sea* (1975); *Dokyumento '75—Minamatabyō to Kanada Indian* (Document '75: Canadian Indians and Minamata Disease) (1975); *Message from Minamata to the World* (1976); *Minamata: Those 20 Years* (1976); My *Town, My Youth* (1978); *Umi to otsukisama tachi* (The Sea and the Moons) (1980); *The Minamata Mural* (1981); *Minamata: Those 30 Years* (1987); *Memories of Kawamoto Teruo—Minamata: The Person Who Dug the Well* (1999); and *Minamata Diary: Visiting Resurrected Souls* (2004). Siglo and Independent Films have released a seventeen-DVD set of Minamata-related documentaries, including those by Tsuchimoto Noriaki, as well as Ichinose Masafumi, Koike Hiroko, Koike Masato, Onitsuka Iwao, and Satō Makoto.

the state" (Tsuchimoto, 2004a, 47). This position, which started at the bottom and looked up, aimed to overturn the layers of silence and exclusion that surrounded Minamata Disease. From that position the films argue for an appreciation of knowledge as something emergent and egalitarian, something that is never complete but always evolving over time, and something that everyone is equally party to. If official knowledge and social prejudice had enabled the decade of silence, against this the films present an emergent and egalitarian form of knowledge as something that is both truer and more humane.

Although more than fifty years have passed since the first official discovery of Minamata Disease the struggle against the ignorance and silence surrounding it has still not ended. Minamata may be well known as one of the most disastrous cases of mercury poisoning in history, but it is not yet properly a part of history. As of 2010 no clear definition of Minamata Disease exists in Japan and no comprehensive health survey has been conducted of the area exposed to Chisso's pollution. There is still no way of knowing how many people living around the Shiranui Sea may have been affected (Higashijima, 2010, 9–13, 240–43). The official stance continues to be characterized by hostility toward the work of local researchers and a refusal to fund the kind of comprehensive research needed to understand the full scale of the damage. The question of knowledge, and who constitutes it, is therefore still very much alive for the people living around the Shiranui Sea.

## HISTORY OF THE MINAMATA POLLUTION CASE

Perhaps emblematic of the lack of comprehensive research surrounding Minamata, the cause for the massive mercury discharge from Chisso's Minamata plant was not discovered until 1998. Although the plant had been using mercury for acetaldehyde production since 1932 one of the long-running mysteries of the case was why the outbreak of Minamata Disease did not begin until two decades later, in the 1950s. The reason for the lag was finally spelled out by researchers working with an insider who had worked at the plant. Their research showed that in August 1951 the plant had changed the oxidizer used in the production process from manganese oxide to nitric acid, greatly increasing the amount of organic mercury expelled as waste. Within two years of the innovation the amount of organic mercury in the factory's wastewater rose from under 10 kg per year to more than 50. By the end of the 1950s the level was approaching 100 kg per year (George, 2001, 69–70). This single innovation was the mechanism that caused the pollution, but the reasons that its effects would grow to such enormous scale are more complex.

What would become the Chisso chemical plant was built in Minamata in 1908.* At that time it single-handedly saved a town facing extinction due to the decline of its traditional salt-producing industry. For much of the following century the Chisso plant would be identified with the town's fortunes. It brought the best and the

* I use the name Chisso throughout for convenience's sake. The company's name was Nitchitsu from 1908 until it was broken up by the Allied Occupation after 1945, and reconstituted as Shin Nitchitsu in 1950. It changed its name to Chisso in 1965 (George, 2001, 31–33, 163).

brightest in science and engineering to Minamata, and the company was "a leader and model for Japan's chemical industry, just as Sony and Honda became models for postwar technological entrepreneurs" (George, 2001, 18). The plant connected what would otherwise be a geographically isolated region to Japan's modern development, and for Minamata the plant was the only such connection.

The modernization that Chisso brought with it did not end long-surviving forms of domination and discrimination, however. When company executives and engineers arrived they set up their new homes in a neighborhood long favored by the old elites, still located at the foot of where Minamata's feudal castle had once stood. The identification was so complete that Minamata was often referred to, only half-jokingly, as Chisso's "castle town." While the company executives settled easily among the traditional elites at the pinnacle of the local hierarchy, the factory managers were on the next rung down, followed by workers at the plant. Next were the townspeople who served the needs of the company and new residents and became increasingly dependent upon them. Lower still were those who lived in the outlying areas: the fishermen and farmers who did dirty, difficult, and distinctly unmodern work. Of these the farmers at least could lay claim to some cultural capital that came with rice-growing, something unavailable to the fishermen. The fishermen were, therefore, already among the most marginalized populations in Minamata when they were struck with the full impact of Chisso's mercury pollution. The effects of the pollution itself only deepened their marginalization, turning it into outright ostracism as rumors spread that the unidentified "strange disease" (*kibyō*) was contagious or the result of bad breeding.

The effects of the mercury appeared in animals first. In the early 1950s, barnacles began to fall off the fishermen's boats and fish could be seen swimming so slowly that they could be caught by hand. These were soon followed by the fishermen's cats, which began to behave erratically and die en masse. By the mid-1950s the effects had appeared in humans: loss of feeling in the extremities, impaired coordination, slurred speech, constricted field of vision, convulsions, and eventually a slow, wasting death. The local health department declared the appearance of a new disease in 1956 and over the next few years, scientists, mostly based at local and regional institutions, zeroed in on the cause of the disease. It was well known that Chisso used inorganic mercury to produce acetaldehyde. By 1959 a group of researchers at Kumamoto University had shown that extremely high levels of organic mercury were present in the sludge outside the factory's waste outlet. They also had shown that organic mercury was present in the local fish and shellfish, and that it produced the same symptoms as Minamata Disease when fed to lab animals. The only missing link was to show how the inorganic mercury used in the process became organic mercury in the sludge (something they established in 1961) (George, 2001, 56–61).

People inside Chisso already knew that organic mercury was being produced inside the plant. As of 1959 their own internal research had also shown conclusively that the wastewater from their acetaldehyde production process produced the symptoms of Minamata Disease in cats. But this research was quashed and any information about the process jealously guarded. Chisso embarked on a public relations blitz to discredit all theories that pointed toward mercury, showering the media with possible counter scenarios ranging from fertilizer runoff to mysterious World War II ordinance supposedly dumped in Minamata Bay but never confirmed to exist. Government was complicit in obstructing research that pointed toward Chisso's mercury. Immediately following the submission of the Kumamoto University team's report in 1959, the Ministry of Health and Welfare disbanded the team and ended its funding. The findings of the report were buried. Continued Minamata Disease-related research at the university was possible only thanks to a US National Institutes of Health grant that ended in 1963. George concludes that "in retrospect, these events are discouraging. Too many scientists seem to have been in the service of money and power. Too many in the media saw it as their duty to be 'neutral' by uncritically reporting every theory. ... Too many government officials seem to have been willing to sacrifice poor fisherfolk on the altar of high growth" (George, 2001, 70).

If industry and government succeeded in confusing the message, there was also another, parallel problem: there was no clear audience for the message at the time. In 1960 Chisso made up 48% of Minamata's tax receipts and together with its subsidiary Shin Nihon Kagaku employed 19.2% of its population (George, 2001, 35). When the fishermen and pollution victims began to demand compensation from Chisso in 1959 they not only spoke from the social margins but were attacking the very institution that most residents saw as the town's lifeblood. When in 1959 some sort of punitive action against Chisso looked possible, the mayor and the city council, together with the Chisso labor union, train workers' union, teachers' union, small manufacturers' union, and taxi companies all voiced their opposition to action against the company (George, 2001, 95–96). The victims of the mercury pollution had no natural audience in Minamata and had been unable to reach a wider audience. In late 1959 they were compensated with a solatium payment (which recognized suffering but did not admit responsibility), and in 1960 Chisso installed some impressive-looking filtering equipment in a wholly cynical and successful bid to allay public fears about ongoing pollution. With that, virtually everyone in government, industry, and among the public considered the case closed. Over the next eight years Chisso continued to dump untreated organic mercury into the Shiranui Sea, fishing from the sea continued, the public remained ignorant, and the isolated fishermen and disease sufferers tried to cope with the damage to their lives and livelihoods with their meager compensation.

There are a number of reasons why the next round of struggle, which began in the late 1960s, turned out differently. Arguably the single most important difference was that in this second struggle the victims were able to forge alliances beyond the confines of the local power structure. They made contact with another group of victims of mercury pollution in Niigata, and following that group's example some of them filed a lawsuit against Chisso in 1969. They began to find sympathy among a wider public, as well as material and spiritual support from a growing network of activists around Japan. Their struggle became known well beyond Minamata thanks to the work of people who helped communicate it. In 1968 scientist and activist Ui Jun published his damning study of the disease and its cover-up titled *The Politics of Pollution: On the Trail of Minamata Disease*. The next year, Ishimure Michiko, writer and longtime advocate for the disease victims, published a powerfully empathetic account of their lives in *Paradise in the Sea of Sorrow: Our Minamata Disease*.* A few years later photographers W. Eugene and Aileen M. Smith arrived in Minamata and their work would bring images of the victims to an international audience for the first time. Tsuchimoto Noriaki first joined the Minamata victims' movement in 1970. He participated in a protest by Minamata victims at Chisso's Tokyo headquarters and was arrested. This put him in contact with members of the movement and with the help of an introduction from Ishimure Michiko, he arrived in Minamata that summer to begin filming (Tsuchimoto, 2004a, 69–70).

## *Minamata: The Victims and Their World*

*Minamata: The Victims and Their World*, released in 1971, was Tsuchimoto's first major documentary about Minamata, and it remains the best known.† As the title suggests the film introduces the victims of Minamata's mercury poisoning and the world they inhabit. Most live in difficult circumstances, suffering from mercury-related disability or having lost relatives in the disaster. We meet them person by person, family by family, in a series of portraits where the people speak about their experiences and the camera shows them in their everyday surroundings, sitting at home, fixing nets, and working out at sea. Their testimony fills the soundtrack. Apart from a few doctors and the occasional interjection of Tsuchimoto himself, the film cites no experts: The voices and narration are the words of the patients themselves as they talk about their struggles. An older man remembers how, when his daughter died in the late 1950s, he had to carry her body home from the hospital on his back because no taxis would take him. A young woman tells how she was forced to divorce her beloved husband when his family found out she had the disease. A fisherwoman reminisces about how much her father knew about the sea and the weather, saying she's lost without his knowledge. By presenting the testimony of so many of the victims the film gives voice to those who for so long had been discounted for the sake of developmental goals. These testimonies would now be part of public knowledge so that the calculus of high economic growth could not be the same.

There is also a narrative to the film: that of the victims' struggle for recognition and compensation. The film concentrates mostly on the activities of the "trial group," so named because they had chosen to sue Chisso for compensation rather than leaving it up to a mediation panel as the other main block of victims had. The "trial group" was the most politically confrontational of the victims' groups. The main narrative arc follows the birth and progress of the group's "one-share movement," in which they buy Chisso stock certificates so that they can attend a semiannual shareholder meeting and confront company management. The unforgettable climax

* Ishimure's book has been translated into English by Livia Monnet (Center for Japanese Studies, Universty of Michigan, 2003). Although Ui Jun's book has not been translated, he is contributing editor of the English-language volume *Industrial Pollution in Japan* (Tokyo: United Nations University Press, 1992), available online at http://www.unu.edu/unupress/unupbooks/uu35ie/uu35ie00.htm#Contents.

† In 2011 Zakka Films, in coordination with Siglo and Cine Associe, will release a four-DVD set of Tsuchimoto's films with English subtitles. The set will include *Minamata: The Victims and Their World* (1971), *On the Road: A Document* (1964), *Traces: The Kabul Museum* (1988) and *Another Afghanistan: Kabul Diary* (1989).

comes when Hamamoto Fumiyo confronts the president of Chisso with the tragedy of losing both her parents. She screams into the president's face that no amount of money is enough, that all she wants is her parents back. The confrontation is the culmination of the one-share movement, but at the same time serves as the emotional culmination of all of the testimonials, which point to a problem that lies beyond restitution. *Minamata: The Victims and Their World* is undeniably a social-movement film but it also indicates the limits of political redress. Significantly, the film does not end with the confrontation but takes us back to Minamata for a short closing segment, where we see the fishing boats going out again, continuing their daily round of work. The political movement is important but the victims must always return to their world, and the film is, at its core, an attempt to portray that world.

Let us look more closely at the portrayal. Viewers of this and other Tsuchimoto documentaries may be struck by their length and loose structure. They do not satisfy the hunger for knowledge quickly. Rather than a single, authoritative argument about the victims and their world the films present a certain way of understanding that world, one that is committed to the principle that knowledge is never settled, that it is something always being built and rebuilt. In the case of Minamata, when Tsuchimoto and his crew arrived on location to shoot, they did not have any script. The film was unplanned from the beginning, its content and narrative growing out of the experiences of the filmmakers over the five months of their shooting. The most important enabler of this practice was a position among the patients—a level of trust that would allow access to their world. Without this trust the "reverse view" that Tsuchimoto hoped to present would not have been possible. Tsuchimoto knew this all too well from his first attempt at filming the victims of Minamata. The 1965 television documentary, *Minamata no ko wa ikite iru* (*The Children of Minamata Live On*), was work he took on assignment: He did not have any knowledge of the patients' situation or acquaintances within their community when he filmed it. In later writings, he would often recall with chagrin how he was severely reprimanded by the mother of a child with congenital Minamata Disease whom he had tried to film.

In 1970, however, the crew of *Minamata: The Victims and Their World* were invited to live in one of the small fishing hamlets hardest hit by the disaster, at the family home of the above-mentioned Hamamoto Fumiyo. Both of Fumiyo's parents had died of acute mercury poisoning in the 1950s, leaving her with the house to herself. Both she and her brother, Tsuginori, suffered the effects of mercury and were active in the victims' movement against Chisso. The homestead had become much used as a meeting place for those movements. Tsuchimoto and the crew slept and bathed there, shared meals with Fumiyo, and after a day of shooting would view the rushes there together with anyone from the neighborhood who cared to drop in. From that starting point the crew met other families and went to film those who invited them. Eventually they filmed forty-six victims or their relatives and twenty-two children with congenital mercury poisoning. The film's structure reflects this method: the order that the portraits and events appear in is the same order that they were shot in. The one-share movement, which provides the narrative drive, only happened to be hatched by chance during the filming (Tsuchimoto, 2004a, 36–40, 61–76; Tsuchimoto and Ishizaka, 2008, 137–59).

Tsuchimoto and his crew were part of a generation of politically active and formally experimental filmmakers pushing the boundaries of independent documentary in the 1960s and 1970s. Before *Minamata*, Tsuchimoto had directed films about the harrowing work of Tokyo taxi drivers, the discriminatory treatment of a Malaysian student at a Japanese university, and a small group of radical students and the young professor who led them during the violent student uprisings of the late 1960s. His work continually drew him to marginalized subjects and political movements. In addition to these political commitments Tsuchimoto's generation is characterized by a particular approach to filmmaking, which saw documentary film as something that emerged from the relationship between filmmakers and filmed subjects (Nornes, 2002 and 2007). In the West, filmmakers and film theorists have often theorized documentary as an authorial statement: films are understood to document the way the filmmaker sees the world. The idea that a film can provide a space for the subjective empowerment of the people being filmed is treated with suspicion because the filmmaker ultimately has overwhelming authorial control. This generation of Japanese filmmakers, however, made it part of their practice to renounce such authorial control and explore the possibility of making films that were not solely the work of the author looking at the world, but were generated intersubjectively between the filmmakers and the people being filmed in the specific time and place of the shooting. One of Tsuchimoto's juniors remembers this practice as "symbiotic" (Suzuki, 1993, 14), and the ecological metaphor is apt. His idea was not that filmmakers would melt into the world of the subjects to become one with them; the filmmaker holding the camera would always be a different kind of entity. But while the people filmed and the film crew were irreducibly different entities, their meeting and interaction produced a meaning of an order different from any one of them. It was that meaning that the documentary film preserved. It is this idea that lies behind Tsuchimoto's oft-used phrase, "film is a work of living things (*eiga wa ikimono no shigoto de aru*)."* Rather than in the authorial/technical procedures of cutting and splicing, film's work took place in the specific relationships and interactions among living things.

This approach to filmmaking made possible an intimate and highly detailed portrait of the daily life of the fishermen and victims. We see them at home, at work, and in a variety of venues as they continue to get on with their lives. The lives of the "victims," therefore, are not defined by their tragedy. The filming invites us to appreciate the complexity of their continuing way of life and the sensitive understanding of the environment that it rests upon. One sequence introduces us to a family preparing bait. The camera doesn't focus much on the people, but on the bait itself being prepared: hands drop in the ingredients and a giant ladle stirs it in the cauldron. Later, when the mixture has cooled, a chorus of busy hands molds it into shape for the hook. A man's voice explains that each family has its own secrets, but he swears by his because the fish love it—it's delicious, nutritious, fit for human consumption. Anyone can catch a fish that's already there, he confides, the real skill lies in getting the fish to come to you. This scene portrays an intimate and highly attuned form of knowledge, one where the family's own life is sustained by understanding how fish behave as living things.

---

* This phrase would become the title for Tsuchimoto's first collection of essays, published in 1974.

In another scene the camera follows a man wading in chest-deep water, catching octopus. The sequence is poetic, set to gentle music, complete with underwater shots of the octopus tentacles waving in the currents. On the soundtrack the man explains how he lures them out, snags them on his hooked pole, and how he kills them by biting them in a particular "vital spot" between their eyes. As his voice explains the technique the camera observes in close-up as he demonstrates, bringing a squirming octopus up to his face and biting, essentially kissing it in order to kill it. The scene provides such a memorable image that it was used for one of the film posters in the early 1970s (Tsuchimoto and Ishizaka, 2008, 152). In this and similar scenes we understand the visceral and specific link between the fisher families and their environment. They must learn the way the sea works as a thing in order to be able to live off it. Their work is finely adapted to the world they live in, as much a "sense" or "feel" for how things should be done as something that can be expressed in words.

The film gives great attention to these small things of the world, the everyday ideas and habits by which the fishermen connect with the world around them. The camerawork of Otsu Koshiro is instrumental in this: his extreme close-ups and remarkable use of deep space make the viewer consciously aware of the connections between the subjects and their world. The connections being enacted in one's daily life usually become so accustomed that they dissolve into invisibility, but the filming brings the substance and grammar of these connections into the foreground, lingering on them in their physical minutia. The relationships with the world thus portrayed are terribly intimate and sensitive, but in this finely balanced interdependence they also become vulnerable.

While the film is interested in the intelligence embodied in the relation of the fishermen to their environment, it is equally interested in how the more severely affected victims of mercury poisoning get on. One of the families that was hardest hit by the pollution was the Watanabes. The grandmother died of acute poisoning, and all three of the grandsons were born with the congenital form of the disease. *Minamata* shows them at home. One of the young grandsons is playing the organ while his grandfather watches him, and his father sits on the veranda repairing a net. The filming of the family parallels the grandsons' bodies with those of the healthy father and grandfather. An extended close-up of the hands and fingers of the boy playing the organ is followed by a similar examination of the grandfather's knotted and wrinkled arms. The camera lingers over the handiwork of the father repairing the net and then, finally, frames the grandfather, father, and two boys as they hold their hands out in a row. The camera moves over the row of hands twice revealing few noticeable differences (see Figure 8.1).

More important than the paralleling of physical appearance is the kinship that is established between the activity of the disease victims in building connections with their environment and that of more healthy people. The elder boy plays the organ and the younger, though deaf and mute, enjoys "listening" to music with his fingers. In the scene following the one above, the two sons are filmed together in the living room, listening to music at high volume. The older son conducts in time with the music, while the younger puts his fingertips up to the speaker so that he can feel the sound. He puts his hand on the speaker, withdraws, touches his lips, puts his hand back, withdraws it again, shakes it, and again puts it back on the speaker. The camera concentrates on each of these motions before coming to rest on a close-up

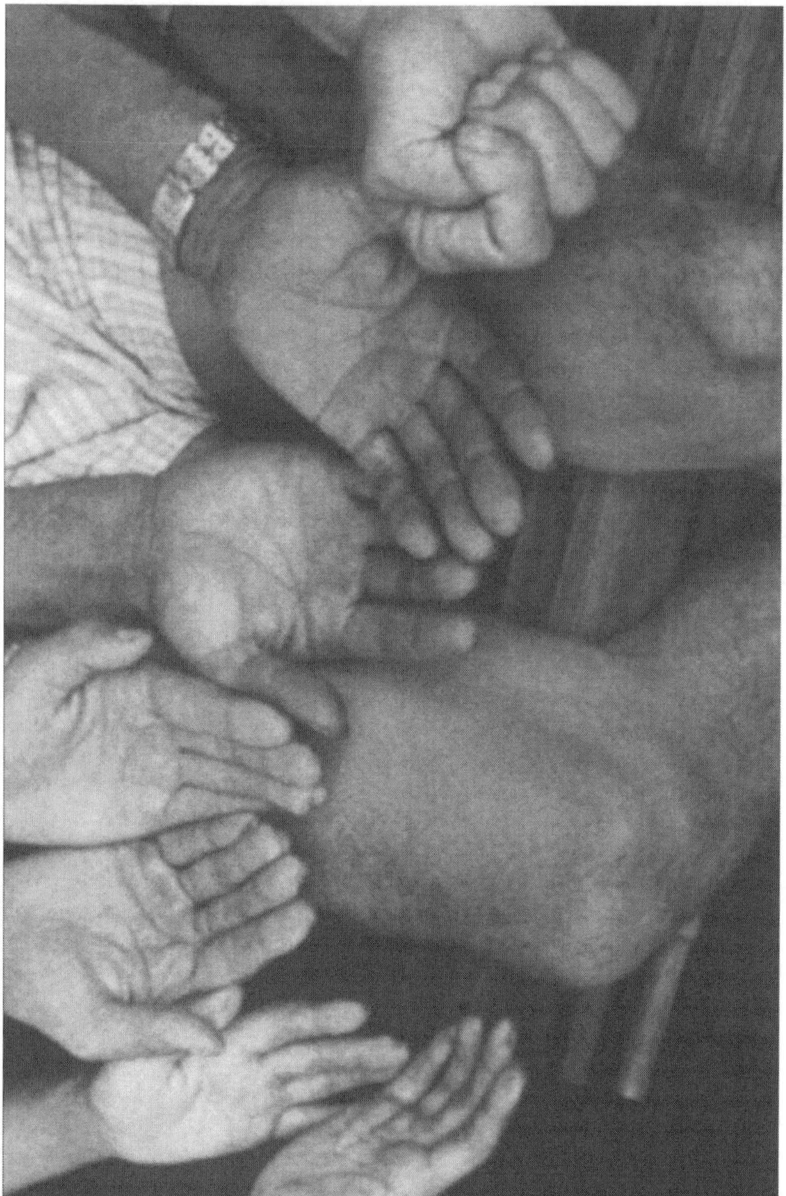

**FIGURE 8.1** From *Minamata: The Victims and Their World*. Courtesy of Siglo Ltd.

of his head. Only at this point does the audience become aware of the deformity of the boy's ear. In affording such time and attention to their activities, recording them gesture by gesture with as much seriousness as the father fixing his net, the film invites us to consider how the boys are also connecting, communicating, and finding meaning in their world. Although the connection might be frail the film suggests that indeed all such connections are. All of the habits that humans and communities develop in connection with the world around them, from fixing nets to finding fish, rest on a fine thread of understanding and in that they are vulnerable.

The shadow of vulnerability hangs over the entire film. It is something that we cannot see directly but stays off camera as memory or potential. The film celebrates the ingenuity and spirit of the victims and fishermen in their relationship with the world but it never lets the audience forget that that relationship has been forever altered by pollution. The alteration is not one that entails an obviously visible degradation of their lived environment. It is an invisible presence that introduces a permanent potential for disconnection, a pervasive disenfranchisement. In one fleeting example a man who had once been incapacitated by mercury demonstrates how he is finally able to hold a glass of water again after years of rehabilitation. The camera frames his face and hand in turn and then there is an unusual shot, which keeps the shaking cup of water in the foreground while looking up the man's arm to his face (see Figure 8.2).

On the one hand we have a great achievement. The old man is able to connect with his hand and establish the sensory feedback loop that mercury had destroyed. At the same time this relationship is called into question. The hand and head are separated from each other, their distance emphasized. Holding a cup of water is one of the most trivial forms of understanding. It might as well be invisible. But the film shows how, after the poisoning, even this most banal relationship is no longer

**FIGURE 8.2** From *Minamata: The Victims and Their World.* Courtesy of Siglo Ltd.

stable or given. The film never tires of documenting the small marvels of daily life, but at the same time calls them all into question, bringing home the sheer scale of this disaster.

While indicating the scale of mercury's effects the film does not make a strong claim that we can fully understand the suffering they cause. The weighty and emotionally difficult testimony of the victims recorded on the soundtrack is often oddly incommensurable with the images of everyday life that the camera gives us. One woman tells the story of her young daughter's death and how the autopsy revealed extensive brain damage that was the result of mercury poisoning. As she does so the camera concentrates on her hands as she uses an ashtray on the floor in front of her to illustrate which portion of her daughter's brain was destroyed. In another scene the grandmother of a boy with cognitive and physical deficits from being poisoned *in utero* relates how the boy sleeps most of the day, doesn't eat much, and needs help going to the bathroom. Unconcerned by this the boy himself reaches out with his hand and begins to play with the camera while it's filming. There is a gap between the experiences as related by the victims and the comparatively banal idioms of everyday life, a gap that avoids both grandiosity and pathos, and makes it difficult for the viewer to identify the tragedy directly with the world of the patients as presented.* The tragedy always remains beyond view at a scale that seems to be both much larger and much smaller than what we can readily imagine. The great achievement of the film lies here, in showing how the tragedy unfolds at this intimate level every day, and by showing that although we might never fully understand it, it is here, at this level, that we must begin to attend.

## THE *MINAMATA DISEASE* TRILOGY AND *SHIRANUI SEA*

The achievement of *Minamata: The Victims and Their World* was also its shortcoming. While it did a remarkable job of showing the victims and their world from a position within it, it contained relatively little information about the mechanism of mercury poisoning and its wider effects on the environment. Tsuchimoto heard this critique many times when he was screening *Minamata* around Japan and Europe in 1972. After returning from Europe Tsuchimoto and his producer, Takagi Ryūtarō, resolved to make three one-hour documentaries about the environmental and medical science of mercury (Tsuchimoto, 2004b, 18–29).

This effort resulted eventually in four films totaling 420 minutes. Although these films lack much of the formal uniqueness and explosive political confrontation of *Minamata*, they might be considered a more remarkable achievement. Remarkable first for the depth of research they required. The crew would spend far longer researching and shooting these films than the five months that they spent on *Minamata*, and the result is probably the most sustained explanation of the effects of

* This gap is created in part by the lack of synchronized sound, which the filmmakers undoubtedly encountered as a regrettable limitation of the technology. Their slightly older camera (an Arriflex 16ST) was not built for sync sound and made so much noise that they couldn't record sound if it were running nearby. The sound recording had to be done before, after, or at a safe distance from the filming, so the cameraman attempted to avoid shots of people's faces as they spoke (Tsuchimoto and Ishizaka, 2008, 140–41).

mercury pollution on film. Remarkable also in that they show Tsuchimoto moving out of his comfort zone. No one in the crew had expertise in medicine or science, and because of the politically-fraught nature of the science involved it was a controversial field. Stylistically also, the films began to adopt many of the conventions of the expository documentary such as voiceover narration, diagrams and visual aids, and the use of interviews with experts. This style went against the grain of independent documentary as outlined above, but Tsuchimoto and his crew began using them in order to produce the films that they believed the world needed. The films are remarkable, finally, because they make a unique statement about science. For many of the intellectuals who supported the Minamata patients in the 1970s science was inextricably bound up with the extractive and destructive habits of industrial modernity. Science was one of the enemies. These films, however, demonstrate a more nuanced view in which the problem is not science itself but the way that science so often intersects with state power as a tool to disenfranchise local knowledge. They document how local researchers had been, and still were, on the forefront of knowledge about the disaster, and how industry, state, and the scientific establishment continued to marginalize their work in order to avoid fully acknowledging the extent of the pollution.

In 1973 the victims of Minamata won their legal fight in a landmark court decision that established a framework for compensation. But their struggle did not end there. The court decision and subsequent negotiations laid out guidelines for reparation for victims of Minamata Disease, but a critical question remained: who has Minamata Disease? Only those officially certified through the government-administered certification process qualified for reparation and other forms of support. Though a certification board had been in place since 1959, after the 1973 court decision it became the focus of increasing controversy. It was cumbersome and inscrutable and often seemed designed to exclude possible victims rather than help them. By 1975 there were already 3,359 people who had applied but had their applications rejected or the decision deferred. By 1980 this number had risen to 7,704, and by 1985 there were 11,094 (Tsuchimoto, 1987). Patients' groups launched a series of court cases over certification in the 1970s and '80s, most of which were found against the government. A political decision brokered in 1995 was widely advertised as the final solution to the continuing legal controversy surrounding Minamata, but as of 2010 the controversy continues and the number of people who are waiting in the medico-legal limbo of deferment has passed 6,000—its highest point ever in the history of the disease (Higashijima, 2010, 11).

The *Minamata Disease* trilogy, completed in 1974–75, addressed the early controversies surrounding certification. It presented a critique of the official science underlying the certification process, arguing for a holistic approach to diagnosis and one that acknowledged that the science of Minamata Disease was not yet well understood. Tsuchimoto interviewed dozens of scientists and doctors and followed some of them in their research. Their research is presented as an ongoing process of discovery in the face of the unknown. The implicit argument against official science was that it was more committed to enabling government policy than expanding knowledge through research. Official science, as constituted in the certification boards, approached the disease as something residing in a clear, discrete set of symptoms in an individual body, while Tsuchimoto's films showed that the disease could not be so easily defined: many people expressed only partial symptoms, and accurate diagnosis had to incorporate the person's environment and lifestyle together with their symptoms. As Tsuchimoto would write Minamata Disease was a social disease, not only in its causes but in the ongoing battles over its very definition (Tsuchimoto, 2004b, 20).

The three installments in the *Minamata Disease* trilogy are *Progress of Research*, *Pathology and Symptoms*, and *Clinical Field Studies*, which Tsuchimoto glossed as presenting the "history, theory and practice" of the science of Minamata Disease (Tsuchimoto, 2004b, 79). *Progress of Research* introduces the research done on the disease to date, concentrating heavily on the initial phase in the late 1950s before the cause and nature of the disease were known. The film is mostly interviews in which the scientists and doctors who worked on the early research relate the process of trial and error. The structure succeeds in bringing the viewer back to the time when there was no established knowledge to rely on. Minamata was the first scientifically documented case of mercury poisoning through the food chain so the scientists' work was fundamental research. They explain to Tsuchimoto and the viewer what information they had had and how they used it to rule out certain possibilities, gradually homing in on mercury as the cause and beginning to understand how mercury worked on the body. To help explain their work the scientists use visual aids such as diagrams, maps, and archival films, which are presented to the viewer in formats that sometimes mimic lectures, sometimes intimate conversations. Rather than presenting a single filmic narrative about Minamata Disease this film is a series of scientists and doctors who themselves present their research. This structure was necessitated in part by the filmmakers' lack of expertise, but it also shares the basic approach taken in *Minamata: The Victims and Their World*, in which both filmmaker and film take the position of learner.

In the second installment, *Pathology and Symptoms*, science becomes a matter of considerable technical expertise. As in the previous film, knowledge is imparted to the viewer by scientists using visual props to explain how mercury attacks the body. Also like the previous film the viewer is presented not only with the conclusion, but with a detailed explanation of the processes of thought and experiment that led to the conclusion. One long segment explains how monkeys were fed mercury, frozen, and then cut into cross sections in order for scientists to understand where mercury accumulated in the body. In addition to providing a great deal of information about the disease, however, the film also reveals the limits of a purely technical approach. Definitive diagnosis was the province of pathology, which suffers from the limitation that the organism must be dead for a conclusive study to begin. Official certification encountered similar absurdities in demanding that an inexact science be made an exact one. The film introduces what was taken as the gold standard of the official diagnosis: an eyesight test that could detect small breaks in the eye's ability to track a pendulum swinging in front of the patient. The government favored this test because it was impossible to fake and it provided a simple, black-and-white answer. But the test had many problems. Patients might not exhibit the symptom, while others might not be able to see well enough to complete it. The test required the patient to be able to sit upright, concentrate on the pendulum, and communicate with the doctor. Tsuchimoto interviews one family whose son's cognitive and muscular impairments made it impossible for him to sit through the test. When Tsuchimoto brought this young man's case to the Kagoshima City office to confront a local official about it, the official says in effect that there is no way around it. Thus, the official need for technical precision and definitive answers is shown as being unable to account for actuality.

This problem, posed at the end of the second film, leads into the final installment of the trilogy. Unlike the unidentified voiceover narrator of the other two films, the final film is narrated by Dr. Harada Masazumi. Harada is well known as a doctor who spent a great deal of time treating and studying the victims of Minamata, possibly more than any other single doctor. He discovered and established congenital Minamata Disease almost single-handedly and has written many books about the disease.* To a large degree this is Harada's film (Tsuchimoto, 2004b, 55). While the camera follows him in his research around the Shiranui Sea he introduces his own theory of the disease and how it must be studied. His first insight is that Minamata Disease is not yet fully understood. He introduces a new and distressing case in which a young man suddenly (literally overnight) developed all the classic symptoms of acute poisoning without any obvious clues in his diet or lifestyle as to why. None of the current accounts of mercury poisoning could adequately explain his case and for Harada this was evidence that the theories should be thrown into question, not the victim.

His understanding is at odds with the official view on other points also: he argues that Minamata Disease affects the whole body (not just the sensory neurons) and that its diagnosis must consider epidemiological factors such as the residence, occupation, and dietary habits of the patient. He takes us to visit families whose members received different designations from the certification board because of small differences in the expression of symptoms. Harada's location, on the ground among the patients and witness to the subtle variety in the way the disease affected people's lives, made it impossible to endorse the black-and-white approach to diagnosis. Across the fabric of behaviors and environments, the disease itself would appear differently and that complex variation could only be understood through extended, ongoing research. It is ultimately Harada's view of science that underlies the trilogy's view as an unending practice of inquiry, which gives priority to the emergent variety of the world rather than the desire to tidy it up. The *Minamata Disease* trilogy demonstrates how the margins, particularly the victims, were excluded from being authors of authoritative knowledge and how, compensation notwithstanding, the same basic system of exclusion that had caused the decade of silence remained unchanged.

*Shiranui Sea* is in some sense the crowning achievement of Tsuchimoto's first five years of engagement with mercury poisoning. The film is a long and powerful meditation on the depth and breadth of the tragedy, and brings together the issues raised so far: the complex integration of natural patterns and human habit in the formation of community and ecosystem, the vulnerability of these systems in the face of pollution, a deep appreciation for the spirit of people coping with adversity, the

* Harada's important 1972 work, *Minamata Disease*, is available in English, translated by Tsushima Sachie and Timothy S. George (Kumamoto: Kumamoto Nichinichi Shinbun Culture & Information Center, 2004).

difficulty of many patients in finding proper recognition, and above all the conviction that the extent of the disease was not yet known. Tsuchimoto extends his area of research beyond Minamata to the entire Shiranui Sea, a body of water about forty miles long and ten miles wide bounded on one side by the mainland and on the other by a string of closely grouped islands. Remarkably this film is the first to give sustained attention to the great natural beauty of this ever-calm inland sea. Shot almost exclusively on brilliant sunny days, it introduces the viewer to a variety of traditional fishing methods, the ingenuity of which is almost as stunning as their setting. It exposes the fact that although fishing had stopped in Minamata Bay it continued basically unchanged on the wider Shiranui Sea. The viewer is treated to shots of the fishermen preparing a feast of fish they have caught, and thanks to Tsuchimoto's unerring empathy this comes across as less shortsighted than it might otherwise. For people whose lives and communities have been shaped around fishing and who have enjoyed a daily bounty of fresh fish since childhood, it is simply impossible to give it up. It is the substance of their work, the fabric that ties them to the world. Yet symptoms of mercury poisoning continue to spread around the sea and further inland. As happened with so many before them the people affected usually go undiagnosed through a combination of their own ignorance about the disease, the incompetence of local doctors, and the social pressures to keep quiet in order to protect the local fishing industry. The same problems, both the health problems and the tensions within communities, continued unabated even as of 1975 when the film was released.

In addition to showing how widespread the disease was geographically, *Shiranui Sea* shows how the effects of the poisoning continue to develop over time. It revisits many of the patients familiar from previous films. The children born with congenital Minamata Disease are growing up. They are now entering their teenage years and with this their life has grown more complicated. Most of the congenital patients now live in a special facility that was built to care for them, but this arrangement leaves them isolated. After a raucous doll festival revue put on by the children for the staff and volunteers, the narrator notes that not a single parent had come to see the show. One boy, who viewers of the previous films will recognize, has begun to develop distinctly adolescent tastes. As one of his caretakers tells us, he is attracted to women, and as comes through more than once his passion for women is matched by a passion for cars. When an older, noncongenital Minamata patient brings his car to the facility, the boy gets behind the wheel, revs the engine, and plays with the shifter. He is so infectiously overwhelmed with excitement that it is a bitter moment when he asks how long it will take to get a license. The audience knows he'll never be able to drive.

Following this there is a scene where a young woman with congenital Minamata Disease has a long conversation with Dr. Harada. The camera is set at a respectful distance so that we can only see her back and hear her voice. She begins by asking Dr. Harada if she can have a brain operation that will make her better. She has begun to realize that she is different from people around her and breaks into tears as she tries to explain how when she looks at things, like the sea or a flower, nothing comes to mind. She knows something should. She sees other children making progress each day at school, but she seems to stay in the same place. It is heart-rending testimony that she can't see a place for herself in the world. The effects of the mercury poisoning continue to be very real in the lives of these young adults, and as they grow up the way it affects their lives will continue to change. Tsuchimoto's legacy in these films is to have shown how these small and hesitant voices are precisely the ones the world must hear.

## SCREENING THE FILMS

In order for those voices to be heard the films had to be viewed. Tsuchimoto was aware that mercury pollution was an ongoing problem and that the people who most needed to see his films might be the ones with least access to them. For him the emergent and egalitarian nature of knowledge was not just a matter of textual statement but of the whole practice of filmmaking and viewing. If the film was something that emerged from an actual, unpredictable process of exploration that both filmmakers and filmed subjects were party to, then the distribution had to be equally cognizant of the fact that the viewers were not an abstract entity but actual people who came to the screening with particular needs. Exhibition and viewing also had to become a work of living things.

*Minamata: The Victims and Their World* was completed in 1971 when the movement supporting the victims in their case against Chisso was reaching its peak in Japan. In most major cities around the country there were groups of activists who published newsletters and held events and fundraisers supporting the victims in their struggle. This grassroots network helped set up screenings of *Minamata* and the film became part of the movement to raise consciousness about environmental issues. Over the four to five years following its release Tsuchimoto estimates that it reached one million people through these groups. Globally, also, the early 1970s was a period when the environment was bursting into the public conscience. In 1972 the United Nations held the first Conference on the Human Environment in Stockholm. Tsuchimoto himself attended together with a contingent of Minamata patients who spoke at the People's Forum section of the conference. Following the conference Tsuchimoto took a three to four month tour of Europe, screening *Minamata* in West Germany, the Netherlands, France, England, and the USSR in a variety of venues (Tsuchimoto 2004a, 139-80).

While *Minamata* was timed perfectly to participate in this global upsurge of concern over the environment, the *Minamata Disease* trilogy and *Shiranui Sea* faced a different landscape. In Japan the group of patients who had sued Chisso won their case in 1973, and as of 1970 the ruling party and bureaucracy had begun to temper economic development goals with protection of the environment and quality of life (Taira, 1993, 172–75). By the time these later films were released, therefore, much of the fervor of the wider environmental movement was dissipating. The films were also longer and more technical than *Minamata*, and contrary to producer Takagi's expectations world governments showed only lukewarm interest (Tsuchimoto, 2000, 70–71).

The dangers of mercury in the environment, however, had not diminished. In December 1973 W. Eugene and Aileen M. Smith received a letter sent to them in Minamata from two private citizens in Canada who were fighting a battle against mercury in the English-Wabigoon River in northern Ontario. Aileen later visited

Canada and was soon writing to people in Japan about it: "It is just unbelievable how this situation parallels Minamata—the same stupidity, the same cover-ups" (Smith and Smith, 1975, 141). The mercury was being dumped upriver by a paper mill that contained a chlor-alkali plant, while downstream there was the same mixture of economic self-interest (the area's major industry was fly-fishing tours), disregard for marginalized populations (the government deemed mercury levels safe enough for tourist sport fishing, but banned commercial fishing, part of the livelihood of the local Cree Nation), and apparent lack of interest in discovering the whole truth about mercury exposure (the first major epidemiological study of the Cree was not completed until 1979). In the face of government inaction some members of the Cree Nation formed an interlocal alliance with the people of Minamata. In spring 1975 a group of Minamata researchers that included Dr. Harada visited some Cree reservations, and in July of that year five members of the Cree Nation traveled to Minamata to learn about mercury poisoning for themselves.

Tsuchimoto accompanied the Cree during their visit, making a short documentary about it for Japanese television. The collaboration laid the groundwork for his own trips to Canada. With the help of local activists Tsuchimoto and a group of assistants made two tours of Native American reservations across Canada, the first from September to December 1975, the second from May to July 1976. Over the course of the two trips they exhibited a combination of Minamata films in more than 110 screenings involving more than 12,000 people.* It is not clear exactly how much this helped the target audience; some lacked the ability to read the English subtitles and there were many empty seats by the end of the screenings. But the screenings were flexible: films could be stopped for translators to fill in and take questions, and as the director of all the films Tsuchimoto had the leeway to reedit them in the process of projection, to concentrate on the segments that seemed the most effective for the audience. After Tsuchimoto returned to Japan he put together a much reduced introduction to Minamata Disease, *Message from Minamata to the World* (1976, produced in cooperation with Radio Quebec), which reedited footage from his previous films based on his experiences with the audiences in Canada.

Tsuchimoto also worked to bring his films to audiences that needed them around Japan: those who lived in the shadow of mercury but lacked the knowledge to protect themselves or seek the redress they might deserve. At the close of *Shiranui Sea* the film crew visits a family that lived in a village on one of the remote islands on the west side of the Shiranui Sea, across from the city of Minamata. The wife of the family in question had died a few years previously and research had shown that mercury levels in her hair were an astounding 920 parts per million (ppm). Tsuchimoto interviews her husband, who describes how she had died. He describes all the classic symptoms of Minamata Disease, but until his meeting with the filmmakers it had never occurred to him that the cause might have been mercury. His wife had never been diagnosed with Minamata Disease.

The city of Minamata is located halfway down the east side of the Shiranui Sea on the mainland coast, and by the mid-1970s the people in the towns up and down

* These numbers refer only to the first, longer trip. Numbers for the second trip are unknown. In Tsuchimoto, *Waga eiga hakken no tabi*, 2000, 87.

this coast were relatively well informed about Minamata Disease. But it was a different story for the people on the small islands across the sea, Amakusa, Goshonoura, Nagashima, many of which were only accessible by boat. The news about Minamata that came to their tiny villages seemed to come from a different world. People saw the pollution problems as a "big city" issue, not something that they had to worry about. The isolation of the communities favored local power holders just as it had in the case of Minamata. There was a local incentive to maintain both silence and ignorance since local fishing cooperatives could not afford to lose their market in a pollution panic. If a resident applied for certification as an official Minamata Disease sufferer they risked intimidation and ostracism. While the Amakusa Islands had been famous during Japan's feudal days for "hidden Christians" who had to practice their faith in utmost secrecy for fear of persecution, they now became known for harboring people with "hidden Minamata Disease" (Tsuchimoto, 2000, 93–258).

Much as they had done in Canada Tsuchimoto and a group of supporters brought the Minamata films on a tour of these islands with the goal of screening them in every village on their coasts. They eventually succeeded, holding screenings almost every day at seventy-six separate locations over the late summer and fall of 1977. Sleeping in local community halls and men's clubs, often holding screenings outdoors, they brought the films to approximately 8,500 people. Although it is difficult to quantify the effects that this tour had there are many anecdotes in Tsuchimoto's personal account of the trip that suggest that the screenings helped people understand that mercury was something that might affect them also; made them realize that loss of sensation, tremors; and birth defects might have causes other than what their doctors told them, and gave courage to people who had considered applying for recognition, or had tried and failed.

## THE CONTINUING STRUGGLE

Tsuchimoto continued to be involved with Minamata for the rest of his life. In the late 1970s he codirected *My Town, My Youth* (1978), which followed a group of young people with congenital Minamata Disease as they organized a public concert featuring the popular female vocalist Ishikawa Sayuri. In 1981 he made *The Minamata Mural* (1981), about the artists Maruki Iri and Toshi, well known from their murals depicting the atomic bombings, as they completed a mural about the tragedy of Minamata. In the 1980s Tsuchimoto's interests expanded: He directed two films about nuclear power and its dangers, Tsuchimoto Noriaki's *Nuclear Scrapbook* (1982) and *Umitori—Robbing the Sea at Shimokita Peninsula* (1984), and an ambitious work on Afghanistan during Soviet occupation, *Afghan Spring* (1989). In his final works, however, he returned to the subject of Minamata.

*Memories of Kawamoto Teruo—Minamata: The Person Who Dug the Well* (1999) used footage from Tsuchimoto's previous films to compose a memorial to the extraordinary activist Kawamoto Teruo, who for nearly forty years had fought tirelessly for the rights of unrecognized disease sufferers. As both Tsuchimoto and key people in the patients' movement neared the end of their lives, this film was also an opportunity to look back over the entire movement. Tsuchimoto's final work, released in 2004, was *Minamata Diary—Visiting Resurrected Souls*. It was shot on video, mostly in 1994–95, when Tsuchimoto and his wife, Motoko, returned to Minamata to carry out a memorial project that was to become part of the landmark Minamata Tokyo Exhibition held in 1996. Tsuchimoto's idea was to display photographs of all those confirmed to have died from Minamata Disease, 1,080 people as of 1994. Eventually the Tsuchimotos succeeded in collecting more than 500 portraits of the deceased, and by all accounts the gazes of these people as they looked out of their funeral portraits at the visitors to the exhibition created a powerful reminder of the human cost of Minamata (George, 2001, 289–90; Tsuchimoto and Ishizaka, 2008, 317).

The task of realizing this exhibition, however, was complicated. When the Tsuchimotos began to gather the photographs they discovered that no public record of the names and addresses of the deceased victims existed. Perversely only Chisso held such a record and they refused to share it, citing concerns for the deceased patients' privacy. The Tsuchimotos had no choice but to track down each family one at a time and ask for their permission to display the photographs. A third of the families refused and some only agreed on condition that they never be displayed in Minamata or Kumamoto (George 2001, 289–90). It took a full year to gather the photos for the exhibition. Although they encountered the long-familiar need to hide the disease during this trip, the extended sojourn gave them the chance to document some of the changes that had taken place in Minamata. The film that eventually resulted is a tribute to the ability of living things to revive and recover. The harbor that had been most polluted by Chisso's effluent had been dredged, much of it transformed into a field of reclaimed land. How to use this land was a matter of some controversy, but *Minamata Diary* records the ceremonies and performances organized by patients' groups that aimed to use the space for public remembering. The film also contains a number of public addresses in which the victims of the pollution speak to audiences about their experiences. With the election of a new mayor, Yoshii Masazumi, in 1994 Minamata City began a number of initiatives to publicly come to terms with the legacies of the disastrous pollution incident. Among the public events Tsuchimoto intersperses poignant observations of the changing of the seasons and the sea life returning to the shores of Minamata Bay. The theme of resilience runs through all of these, and Tsuchimoto's final meditation in his narration is the wish that these practices of remembering and sharing show the way to Minamata's recovery.

It is on this ambivalent note that we must end. Compared to the decade of silence in the 1960s much has been achieved for the victims of Minamata Disease. Many groups and institutions support the victims of mercury poisoning and preserve the memory of what happened. The Minamata Disease Center Sōshisha, founded in 1974 by Ishimure Michiko and other activists, continues to offer support services to patients while maintaining a museum and document archive in a number of languages. In the mid-1990s Minamata City built a Municipal Minamata Disease Museum and a memorial to the victims. Today it bills itself as a model environmental city. The Ministry of the Environment supports a Minamata Disease research center, also located in Minamata City. The widespread silence and ignorance that have shrouded Minamata Disease are therefore becoming things of the past.

At the same time some of the fundamental features of the situation that Tsuchimoto encountered while screening his films around the Shiranui Sea in the mid-1970s have not changed. A comprehensive health survey of the Shiranui Sea region has never been carried out either by the prefectural or national governments. There is still no way of knowing just how many may be suffering from "hidden Minamata Disease" (Higashijima 2010, 240–43). Certification still relies on the patient's initiative in requesting designation from an officially appointed board of physicians. When people do apply for certification they face a system that has long seemed designed to minimize potential liability by denying applications. A Supreme Court ruling in 2004 struck down the government's diagnosis criteria as being too narrow, but six years later it has still not instituted new ones (Higashijima, 2010, 163–65, 202–07). In an attempt to skirt the court ruling and impose yet another final solution on the problem, the government passed in 2009 a "Special Measures Law Concerning the Relief of Minamata Disease Victims and the Solution of the Minamata Disease Problem," which, as with so many of the political agreements of the past, offered a one-time payment in exchange for patients' forsaking future legal action. It also ignored the question of legal culpability. Chisso, now a major producer of components of liquid crystal displays, becomes a major beneficiary under the law, as it provides for the company to be split into two so that the finances of its manufacturing business can be separated from its legal liabilities to Minamata patients (Higashijima, 2010, 228–35; Saitō, 2009, 310–12). For Japan's national government, Minamata Disease is still an occasion for evasion and willful ignorance. If the lessons of Tsuchimoto's documentaries are a humility before the unknown, an awareness of a shared vulnerability, and the need to attend most carefully to the voices that are the faintest, these lessons have yet to be learned.

## REFERENCES

George, Timothy S. 2001. *Minamata: Pollution and the Struggle for Democracy in Postwar Japan.* Cambridge: Harvard University Press.

Harada, Masazumi. 2004. *Minamata Disease.* Translated by Tsushima Sachie and Timothy S. George. Kumamoto: Kumamoto Nichinichi Shinbun Culture & Information Center.

Higashijima, Dai. 2010. *Naze Minamata-byō wa kaiketsu dekinai no ka* (Why Can't the Problem of Minamata Disease Be Solved?). Fukuoka: Genshobō.

Ishimure, Michiko. 2003. *Paradise in the Sea of Sorrow: Our Minamata Disease.* Translated by Livia Monnet. Ann Arbor: Center for Japanese Studies, University of Michigan.

Nornes, Abé Mark. 2002. The postwar documentary trace: Groping in the dark. *positions: east asia culture critique* 10, no. 1:39–78.

Nornes, Abé Mark. 2007. *Forest of Pressure: Ogawa Shinsuke and the Postwar Japanese Documentary.* Minneapolis: Minnesota UP.

Saitō, Hisashi. 2009. *Niigata Minamata Disease.* Translated and edited by Aileen Mioko Smith, Timothy S. George, Kanaya Naomi, Ken Rodgers, Manami Alicia Katagiri, and Mark Palmer. Niigata: Niigata Nippō Jigyōsha.

Smith, W. Eugene, and Aileen M. Smith. 1975. *Minamata: Words and Photographs.* New York: Holt, Rinehart and Winston.

Suzuki, Shiroyasu. 1993. Documentarists of Japan (second in a series): An interview with Suzuki Shiroyasu by Mark Abé Nornes. *Documentary Box,* 2:9–16.

Taira, Koji. 1993. Dialectics of economic growth, national power, and distributive struggles. In *Postwar Japan as History*, Andrew Gordon, ed., 167–186. Berkeley: California Universisy Press.

Tsuchimoto, Noriaki. 1987. *Minamata: Those 30 Years*. DVD.

Tsuchimoto, Noriaki. 2000. *Waga eiga hakken no tabi (Our Journey to Discover Film)*. Ningen no Kiroku 128. Tokyo: Nihon Tosho Sentō.

Tsuchimoto, Noriaki. 2004a. *Eiga wa ikimono no shigoto de aru (Film Is a Work of Living Things)*. New edition. Tokyo: Miraisha.

Tsuchimoto, Noriaki. 2004b. *Gyakkyō no naka no kiroku (Documenting Inside Adversity)*. New edition. Tokyo: Miraisha.

Tsuchimoto, Noriaki and Ishizaka Kenji. 2008. *Dokyumentarī no umi e: Kiroku eiga sakka Tsuchimoto Noriaki to no taiwa (To the Sea of Documentary: Conversations with Documentary Filmmaker Tsuchimoto Noriaki)*. Tokyo: Gendai Shokan.

Ui, Jun. 1992. *Industrial Pollution in Japan*. Tokyo: United Nations University Press. Available online at http://www.unu.edu/unupress/unupbooks/uu35ie/uu35ie00.htm#Contents.

# 9 Mercury-Inspired Arts

*Kira Obolensky and Elizabeth Mead*

## CONTENTS



> [The sense of relation between nature and humanity] in some form has always been the actuating spirit of art.
>
> —**John Dewey,** quoted from *Ethics & the Environment* (Vol. 8)

## INTRODUCTION

When most people think of the metal mercury, they probably think first of thermometers—not art. Yet as Justin Jesty argues in *The Minamata Documentaries of Tsuchimoto Noriaki*, film (itself based on a photographic process that originally used mercury) contributed to exposing the Minamata mercury poisoning and confronted those in power with visible evidence of the tragedy (see Chapter 8). Sydney Plum acknowledges that "the best activist artists represent relationships—between aspects of the natural world, human individuals, and human culture—in such a way as to draw listeners to the cause of sustainable and harmonious lives" (see Chapter 10). Rachel Carson's metaphor of a silent spring bridges a gap between her knowledge as a biologist and nonscientists. Not only film and writing, but painting, sculpture, photography, and performances provide representations of the environment that bring topics, often sequestered in laboratories or board rooms, to the public spaces of theaters, museums, books, and the great outdoors.

In this chapter two artists model ways of artistically responding to environmental pollution in their personal artistic journeys to raise people's consciousnesses about mercury poisoning and our human relationship to the environment.

## *QUICKSILVER*: A PLAY FOR PUPPETS AND ACTORS

### Kira Obolensky

The idea for a play about mercury and the damage it wrought on a hat-making town came to me about fifteen years ago. I found myself visiting small museums, many of them ad hoc, some of them breathtakingly strange. I met a man who collected dust from every country in the world and displayed the glass vials in his garage; I saw a museum of surgery, another dedicated to physical deformities. I remember visiting a museum dedicated to the hat-making industry in a small white clapboard building in Danbury, Connecticut. The photographs of the hatters captured my visual imagination—they looked like wraiths, white and hollow. I read about their dilemma. During the Great Depression, this hat-making town prospered. The men in the wet shops worked until they developed the tremors, the shakes, a form of madness, and then—if they were lucky—they went on vacation to recover. The town's river ran the various colors of the dye lot. A collective madness emerged in that place, a vivid example of madness connected to the bargain we make with prosperity.

But the idea was so logistically big. It seemed the town would have to be a character, and I did not know how to tell this story on stage with a limited number of actors. Also, the idea felt so complex because it combined prosperity, madness, poison, and complicity. An entire town made mad with mercury. I wanted to ask some questions, I felt, in a theatrical way—about how we might escape such a situation, one that feels to me to be incredibly relevant to our modern lives. Was there a way, I wondered, to present the situation in the town, to make the environment itself dramatic? While I'm not sure I began with the idea of creating an ecodrama, I found myself drawn to and compelled by a phrase from the museum that stuck in my head: "We can't live on such strange air."

The idea to use puppets came years after seeing that hat-making museum, after attending a play directed by the great South African visual artist William Kentridge. Kentridge collaborated with the Handspring Puppet Company to create a multimedia *Woyczek on the Highveld*, which I saw at the Public Theatre in New York. That play combined actors and puppets and animations, and it expressed a great scope of landscape and place. I took the idea about hats and madness out of my imagination's museum and began to consider how the story might be told with a combination of song, puppets, visual imagery, and actors. I knew my title would be *Quicksilver*, which to me summons both the mercury that was used in the felting process (removing rabbit fur from pelts) and the idea of speed and fleeting beauty. I wrote the play for puppets and actors. The play has been performed successfully in Minneapolis, where it was produced by 3 Legged Race and The Playwrights' Center in a production directed by Bonnie Schock and myself and then performed in the Prague International Fringe in a shorter, more streamlined version. The first and subsequent productions used puppets made by visual artist Irve Dell, with whom I collaborated; the play was later beautifully performed in Philadelphia by a vibrant young theater company called Gas and Electric Arts under the direction of Lisa Jo Epstein. That production used the script and began anew with visual elements.

In this essay I'd like to address how the play was conceived and my attempts as a playwright to write about the complexity of the issue of mercury poisoning as a fable, with both songs and puppets.

## Organizing an Idea

The first question I had to answer for myself was twofold: why puppets and what would the scope of the play be? I had never written for puppets before. I took a class on puppet-making. I watched other puppet artists, including Michael Sommers, who with Susan Haas runs Open Eye Figure Theatre in Minneapolis. There is a specific theatrical language inherent to the puppet that allows for surreal transformations, communication through gesture, and a grand visual scope on a tabletop. I began to think about how I might populate this town, which I called the Hatmaking Capital of the World. There would be a factory owner, and I wanted him to be a big puppet—with a refrigerator for a stomach. I knew there had to be a hatter on the verge of death. His name would be Gaga Stark (see Figure 9.1). I wanted the hatter to have a family who loved him very much. His wife, Lily Stark, would be a puppet with a spoon for a hand because she would try to bake a magic cake that would save everyone.

The hatter, Gaga, would look as much like the photographs of the hatters I had seen in that museum—he would be white and like a ghost, with a mouthful of crumbling teeth and limbs that were failing him. The dying hatter was built to fly, with a body composed of found wood, and a scissor for an arm that could cut the thread that kept him alive. To make Gaga we looked at the photographs of the hatters I had found through the museum. I made the puppet's heads, first out of clay and then cast them with rubber. I painted them to look white and ghostly. There had to be other characters as well—a doctor who measured misery; a woman who petitions for the workers; and an expert who might tell us about the scientific issues at work. Finally I added a reporter to the mix, a man who comes to town to report on the town's prosperity. This character was inspired by the tone of the newspaper articles of the time I had found through researching the situation—a kind of "gee whiz!" acknowledgement of the terrible situation of the hatters, with a can-do patina that insisted that this terrible situation was somehow vaguely patriotic.

The story would be anchored by three characters who were played by actors: A Man in the Hat, who would play Gaga Stark's soul and serve as our guide; Rose, the young daughter of the hat factory owner; and Gaga and Lily Stark's son, Gregor, who was a poet. At its core I wanted *Quicksilver* to be a play that combined a dilemma with a complex meditation on it; and I wanted to speculate, too, on the ways in which the characters might escape their relationship with the town. Those ways seemed to be death, magic, love, and poetry.

The play begins with an image.

> A puppet, Lily Stark alone in a small chair on the table:
>
> PROLOGUE:
> We start with a puppet. LILY on a chair. By herself.
> A HAT looms over her. A HAT nearly swallows her up.
> It reveals the words:

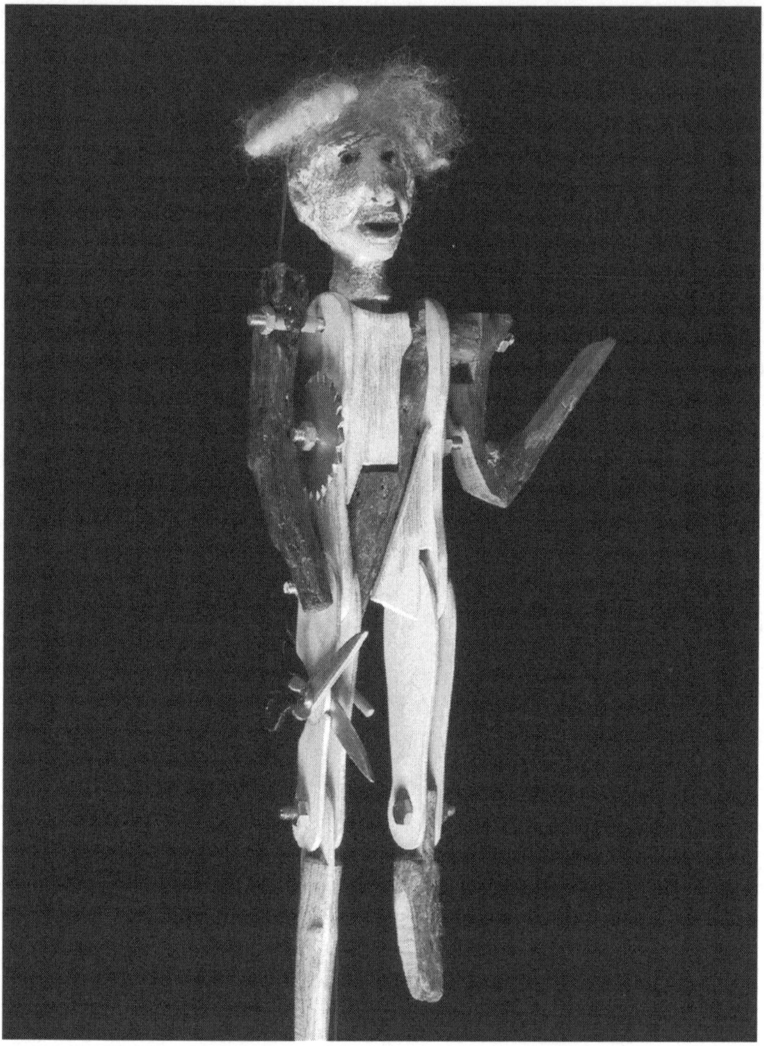

**FIGURE 9.1** *Gaga Stark, Dying Hatter.* Puppet and photo by Kira Obolensky.

"we can't live on such strange air ..."

And a song, revealing the world:

ALL: (sings)
The purpose and function of the hat
is to protect the head
from weather, pigeon shit, and stuff like that
The purpose and function of the hat
is to make the head appear larger
the body less fat.
MAN 1: (sings)

The purpose and function of the hat
is to finish an outfit
to punctuate a suit;

MAN 2: (sings)
to say to the world
I am a person of merit

MAN 1 & 2: (sings)
to accessorize, personalize, fantasize
categorize ...
the purpose and function of the hat

WOMAN: (sings)
The hat has a purpose
and that is to decorate

MAN 2: (sings)
the hat has a function
which might be indelicate
to mention:

MAN 2 & WOMAN: (sung in round)
the hat is a sign
of good breeding, like fine wine, etc. ...

Perhaps the biggest discovery was a visual one. The town would be represented by a long shelf that would hang from above—and perched on it the inanimate puppets would bear witness to the events that would unfold on the tabletop. The shelf created a world that seemed both suspended and awaiting animation—it became the answer to the question, "Why puppets?"

## A Play with Different Points of View

For me the idea of a shared narrative voice in a play allows the theatrical experience to be shaped by different storytellers. This shared narrative function gives the play different points of view, quite literally.

*Quicksilver*'s main narrator is the Man in the Hat, who represents the soul also of the dying hat maker. He introduces us to the town this way:

MAN in the HAT:
Welcome to the hat-making capital of the world.
A pleasant place, where
people make hats but don't wear them much.
This town is a mix
a stew of life,
Polish, Indian, Peruvian, African, Irish
You name it, but we don't.

There are three kinds of people in this town.
The men who work in the wet shops at the hat factory
The women who love the men who work in the wet shops at the hat factory
And the men who own the hat factory.

Each is a little crazy
in their own specific way.
You see, there's a smell,
a steam
a poisonous liquor
that infects the air.
And when you
breathe the sweet smell
of the factory
well,
it turns the brain
into a pot of goo,
But while the rest of the country starves,
I mean we're talking lines for food, shuffling feets through dry dust, horses dead in the road ...
We prosper. As long as we work. And when we get sick,
we starve.
Prosper: you know, jingles in the pockets?
Which makes life ever so more pleasant
in a pleasant berg.

And while the Man in the Hat does serve as the primary tour guide for the play, I wanted the narrator function to be shared by the performers who were operating the other puppets. The Man and the Woman join in the storytelling:

WOMAN:
This is the story of a town
that existed
somewhere
I mean, I think it's in the history books.

MAN 2:
The factory.
The factory's owner lives here in this big house.
This is where Lily and Gaga Stark live. In this little house.
And, the river.
Looks like it's red today.
A red river.

WOMAN:
Shouldn't a river be blue?

With the idea that the performers could speak directly to the audience established, I also wanted to capitalize on the idea that these people could, in fact, play different sides of the coin. The performer who describes the factory owner also gives voice to the puppet of the factory owner. Just as he participates in telling us what is wrong with the town, he also participates in showing us that the factory owner, a self-made man, appreciates the connection between beauty and pleasure. In this scene, Mr. Zaworski, the factory owner, appreciates his appetite (see Figure 9.2).

MR. ZAWORSKI is a puppet who has a small refrigerator for a stomach. He opens it throughout the scene to take out various items of food.

**FIGURE 9.2** *Mr. Zaworski, Factory Owner.* Puppet and photo by Kira Obolensky.

MR. ZAWORSKI:
I'd like a cafe au lait
Or perhaps ... a crème brulee

And after that some smoked eel
garnished with a poached egg.

I can do anything I want
Money is no object.
I've got plenty of loot
I made it through the SMASH
I'm drowning in cash.

Look at the view
through the window.
Quite lovely.
So early, it's still cool.
A new day.

Ahh ... there is such beauty in pleasure.

(ROSE, his daughter, enters.)

ROSE:
What did you say, Papa?

MR. ZAWORSKI:
I said—beauty in pleasure.

ROSE:
Oh.
My party, Papa?

MR. ZAWORSKI:
What about it, sweetheart?

ROSE: (sings)
I've got a pretty pink dress
with silly blue feather
and a cashmere jacket
in case of cool weather...

MR. ZAWORSKI:
Sounds lovely, darling.

ROSE:
The party, you see, I'm wondering about it. I wonder what exactly is there to celebrate?

MR. ZAWORSKI:
You're a young lady. Traditionally, there is a party to welcome a young woman into society.

ROSE:
Do you think I should wear my jacket? Cashmere?

MR. ZAWORSKI:
Absolutely.

ROSE:
Papa, where is the newspaper?

MR. ZAWORSKI:
Why?

ROSE:
I thought I'd read it today.

MR. ZAWORSKI:
Why?

ROSE:
Because I'd like to learn about the rest of the world. And well, now that I'm a young lady—

MR. ZAWORSKI:
The papers, Rose, are written by people. And all people tell stories in different ways. Why if you were to ask three people to describe me sitting in this chair. One would label me imposing. Another aristocratic. The third, perhaps, fat.
There is no need to read the newspaper.

(A MAN in the HAT appears holding his hat.)

ROSE:
Oh! there is a man, Father.

MR. ZAWORSKI:
What man, Rose?
(he turns as the MAN disappears)

MR. ZAWORSKI:
No one there.
Your party—the wicker is painted white. The silver gleams.

ROSE: (sings)
He says it's time to celebrate
the glory of our position
I wonder if we tempt fate

with a party like that
isn't there a hat
in the word hate
I wonder if we tempt fate.

I wanted to show each side of the coin: the prosperous; the perspective of those who were trapped by this untenable bargain—work and illness for a livelihood; and the younger generation, the children of the factory workers and owners who know that they both want something different for themselves. And because the play was performed by only three performers, the lovely echo of doubling, of seeing a person in various roles (puppet and otherwise) would work to help express the complexity of it all.

## Facts and a Fable

One of the challenges in a play like this, which is really at its core a fable, was how to blend the real-life science of mercury poisoning with the tone of the play, which is fantastical. I struggled a lot with balancing this exposition. In early workshops it seemed a matter of finding the right amount of information. Certainly we knew that Gaga Stark was ill—and that he'd been made sick from his work, but the specifics of that seemed difficult to express.

I felt, as I wrote the play, that there needed to be a guest appearance made by an expert who might fill us in on some of the "scientific" issues at hand. So at one point in the play—after we understand that the air is sick, that Gaga Stark is sick— an expert appears.

Scene 8. Enter the expert

(A light is placed to illuminate in a spot light ... an expert. The expert is a finger, the first finger. A small hat is carefully placed on his finger tip.)

EXPERT:
Hello, I am an expert.
Don't you feel better knowing that an expert is talking to you?
Now, a few points of clarification.
One, a good hat makes the man. Why without hats, why we'd be nothing more than wild beasts.
Number two, hats at this time were made from rabbits.

TIGER:
Rabbits!

EXPERT:
Excuse me!?
Rabbit fur.

TIGER:
Rabbits!

EXPERT:
How very peculiar!
Which was removed from the pelts
with the aid of a chemical solvent.
Mercury, mercury nitrate mixed with nitric acid, quite
a deadly little cocktail. Called Carroting, cleverly, because
the chemicals in combination made a lovely orange color.

There's also the earlier reference to the rabbit.

TIGER:
Rabbits!

EXPERT:
Please hold your questions for later!
The solvent was administered through steam
which seeps, which leaks, which travels
like a freight train into the men who stood cooking
in the steam for hours for days for years
turning rabbits into felt into hats... HELP!!

(The EXPERT is eaten by the tiger.)

The Expert is a Brechtian device who intrudes on the action of the play. In each incarnation he is tolerable, it seems to me, because he is funny—manifested as a finger with a little hat on it. He becomes a voice in the play, to my mind, that expresses the source of the poison.

## The Writer as a Character

Two characters in *Quicksilver* are at work writing about the issue at hand. For me this idea that the play would itself examine the ways in which we write about such subjects was of particular interest. I was inspired on one level by the tone of the investigative journalism of the late 1920s—the ways in which the hatters were depicted in newspapers. The tone, the way I took it, was a kind of capitalist patriotism—the idea that people would get sick making your hats inspired one writer to say, "Hats off to the workers."

A reporter, Schwartz, comes to town on this particular day to do a story about the issue.

MAN 1:
A reporter arrives from the Big City.
He thinks he has a scoop!

WOMAN:
A scoop? As in small shovel?

MAN 1:
A story—
about a prosperous place—
the richest spot in the land!

WOMAN:
The richest?

MAN 1:
It's a town that makes the one thing no one
can live without.
Meet Mr. Schwartz, from the Herald.

WOMAN:
Is he smart?

(Remember: Whenever SCHWARTZ talks, it is impossible to understand him. He speaks with conviction, but in an entirely garbled way.)

SCHWARTZ:
A pleasing place ...
A peculiar smell ...

As Schwartz interviews the various people of the town—the factory owner, the factory worker—another character, Gregor Stark—who is the dying hatter's son—tries to write a poem. Gregor the poet is trying to write about the complexities of the time in which he lives. He knows somehow that poetry has the impulse to save him, to help him leave the town he's grown up in and the almost certain job he would take.

GREGOR:
My mother says, "Find a job."
I simply don't know what to do about that.
I'm a poet! What can I do if I'm supposed to be a poet!
I think, frankly, I live in the wrong time.
This may be the very worst time to be a poet.

Gregor Stark is a poet at the beginning of the poet's journey. He struggles throughout the play to find the right words to discuss "Our Time." He wants his poem to get at the complexity of the world he lives in, but he can't find the right words.

(GREGOR tries to write.)

GREGOR:
There comes a time when the earth
is savaged—

That's terrible.
(thinks)

There comes a time, the earth savaged by man.
The earth savaged by man.
The town rift asunder
the world hewn in half.
God. It's too melodramatic.
The world is so immense. How to express immensity?
The world is so infirm. How to express infirmity?

At its core, this question is profoundly personal for me—what do I think writing about such a topic can really do about the issue of mercury poisoning? How do I, as a writer, engage then with subjects that have relevance to the world? For me Gregor Stark's question is apt and one that we all can only attempt to answer: "The world is so immense. How to express immensity?/The world is so infirm. How to express infirmity?"

## MAKING THE ENVIRONMENT

### Elizabeth Mead

> I can't seduce these raucous birds.
> Or sneak up on a willow while they riot there.
> Look—even my shadow's a suspect in this dark.
>
> **—Joanna Rawson, "Wind Camp" 2009**

Several years ago while introducing my sculpture students to modernist works that had so excited me as a young sculptor, I hit a wall. Until now students had gobbled up the excitement of these massive works; the scale and the boldness always thrilled them. But now something major had shifted. I was caught off guard. It is not uncommon for one or two beginning students to look at abstract work and not understand why it is art. I am accustomed to guiding students down that path. However, this response was not by just one or two members of the class but by a good portion of the group. And they were angry at the work I showed. What was going on? As I talked with them, I found they were upset with what they deemed the arrogance of the sculptures. These massive sculptures made of steel took up a lot of space, their construction demanded a lot of resources, and they were permanent. I thought to myself, "Well, yes, all of those things are true; isn't that part of the excitement?" Working on a scale so massive, having so much control and so vast an array of resources at your disposal? The work is going to last forever—what more could an artist ask for? Humility, according to my students, was a good place to start. The artists, they said, had an obligation to be conscious of the impact the materials and processes they used had on the environment. Rather than frivolously consume and produce, they needed to take better care of our resources.

At about this time I had joined with a number of faculty members from different disciplines at my college to look at the problem of mercury poisoning. The Mercury Project, coupled with what I was learning from my students, helped me to develop a course, "Sculpture in the Global Environment: Heavy Metal and the Delta Blues," which explored the relationship between global environmental issues and making sculpture. The course focused on two case studies of environmental catastrophes: the political and social impact of mercury poisoning and Hurricane Katrina. These catastrophes served as the catalysts for creating artistic explorations of and responses to global environmental issues. Slowly I was beginning to see a connection between what my students were telling me in the studio classroom and shifts that had over the last several years been gathering steam in my own studio practice. Their criticism of modernist sculpture got me thinking about the work I was making and its relationship to the environment. I had lost interest in large modernist works years earlier but didn't know why. Yet they contained a wealth of formal information crucial to a young student's vocabulary so I continued to teach them. My thoughts turned to artists who made work with, in, and about the environment. I was torn between the physical excitement of Micheal Heizer's work and the gentle touch of Richard Long's walking pieces. These two had something that the other sculptures did not—they had a specific relationship to their sites, something seriously lacking

**FIGURE 9.3** *Wyoming Landscape 3*, 2005. Ink on paper. 42 inches by 90 inches, by Elizabeth Mead. President's Collection of Art at the Wolf Law Library, College of William & Mary, Williamsburg, Virginia.

in the large-metal modernists' works, which is why they were called "plop art." They were plopped down at their site with little or no regard for the environment or architectural context.

Place, where I am in the world, has always driven my work. It was during a stay at the Jentel Artist Residency in Wyoming that I was able to develop an approach to rendering the landscape. The Jentel Residency allowed me to spend time in a vast landscape devoid of human demarcation. On hikes, often my only reference within that landscape was a herd of antelope or deer. This experience triggered two things, a radical scale shift and a way to depict the landscape that made sense to me. Rather than try to render what I was witnessing literally, I worked on drawings that recreated from memory at the end of each day the experience of being in the landscape (see Figure 9.3).

The way our memory registers spatial relationships tends toward a physical as opposed to purely perceptual knowledge of a space. This difference allows for an experiential response that is separate from perception. Trying to capture the scale of Wyoming on paper would only minimize it. At seven feet wide, my drawings envelope the viewer much like I experienced when hiking in the mountains. Depictions of the landscape vary throughout history and cultures. Each is indicative of given societal attitudes. My drawings are a way of creating a sense of the spaces I experience. They are about my place in the world. The drawings engage space in a literal as well as metaphorical way. I can create a spatial experience without the expenditure of massive materials. The work is quieter.

A drawing, like a photograph, frames the landscape a section at a time, reducing the vastness into digestible bits. As we observe in Nancy Holt's *Sun Tunnels*, Lucin, Utah (1973–76), sculpture also incorporates framing devices as well as time and the landscape. Made of concrete drainage pipes nine feet in diameter, *Sun Tunnels* is aligned with the sun and constellations giving the viewer an ever-changing and evolving perspective of the landscape it frames. In this instance the sculpture exists both as a discreet entity and as an ever-changing form. The interaction of light and its play on the drainage forms as well as the surrounding landscape work to connect object and environment as well as the manmade and the natural. In making an image of the landscape or simply putting a frame on it, we attempt to possess it. The landscape extends all around us, but to render it the artist must impose a beginning as well as an ending. The artist must determine the frame for their work. By containing the land in this way the artist removes it from the continuing whole and lays claim to what has been extracted. Establishing a clearly delineated beginning and end, or starting and stopping point, of the image in front of us makes the landscape manageable. As viewers we can contemplate what is before us because we move from one point to the next and end at the edge of the frame, canvas, or piece of paper.

The eighteenth-century painter Claude Lorrain used a unique device for understanding landscape. The user of the Claude Glass—a handheld mirror positioned over the shoulder—would stand facing away from the landscape in question, which was captured via its reflection. It was thought that the landscape was so astounding in its rapture that it was far too overwhelming to absorb from a frontal view. This way of engaging the landscape served two functions: It allowed the vastness to be delineated into digestible segments and at the same time rendered it diminutive in scale, to be held in one's hands as a metaphorical object. Possessing the landscape as an object is another attempt to conquer it. The mirror changes our relationship to the landscape. If we believe we possess it then it becomes one more "thing" that we own only to be cast aside when we are no longer entertained by it. This opens the door to abuse and neglect.

The desire to possess nature is imbedded in human history. As witnessed in Jennifer Baichwal's provocative 2006 film, *Manufactured Landscapes*, China's construction of a hydroelectric dam to power the country involved the decimation of villages and environmental damage to the Yangtze River. Closer to home, people have conquered the West, irrigated the desert, and lately have even begun to engineer the food we eat. We have duped ourselves into thinking we can control everything; that everything can be consumed and then simply replenished at the snap of a finger. Some claim that the lack of harmony with nature has depleted our spirituality, left us bereft of connection to something beyond ourselves.

Art can be an antidote. Artists reflect, shape, and define culture. They offer a form of spirituality. They remind us of the human element and of our connection to the natural world. Through painting, sculpture, printmaking, and photography, the arts tell us who we are and where our values lie. Art transforms the atmosphere in which we live into something perceived and manageable. It is our mirror to the world. It objectifies and allows us to see; art transforms the environment into landscape, and in so doing it just might be the fulcrum that allows us to find a balance between the natural world and ourselves. Civilization by its very nature builds, uses, consumes. It imposes its presence on the natural world. Our job is to find a way to coexist. We must preserve the environment and at the same time allow civilization to move forward. The difficulty arises because in one sense we are a part of nature, which puts our so-called progress at odds with it. In pursuit of the next invention, the next advancement, we set aside the needs of the land. However, we must work to preserve the environment and to become better stewards of what can so easily slip into oblivion. Art, when it draws us in, makes us aware, heightens our understanding, and shows us the landscape in ways not previously recognized. It allows us to see our emotional connection as well as the intellectual side by side.

Another way of making the environment is through works of art that direct us to particular issues facing the health and well-being of the land. While conversation about the landscape and environment permeates all forms of art, I want to focus on a number of works of art that use a variety of media, methods, and processes to specifically address the issue of mercury poisoning. Works by W. Eugene and Aileen M. Smith, John Knapp, and Linda Wysong address issues of mercury in the environment while adhering to the fundamental formal issues of art. Formal issues compose an object or image; they provide a language to tell a visual story. Balance, tension, rhythm, symmetry, and asymmetry all express qualities that are external to the image shown. For instance, in very broad terms, if a landscape is depicted with round, bulbous forms we tend to think of it as gentle and rolling; if it is described with sharp, angular forms the tendency is to respond as though it is more aggressive. While we tend to look at the individual elements of the image that we can identify, the way that image is depicted gives us the overall feeling or attitude of what an artist is trying to convey. Sometimes the whole is greater than the sum of its parts.

In 1972, photojournalists W. Eugene and Aileen M. Smith traveled to Japan to record and report to the world the horrors of mercury poisoning. From 1932 to 1968 the Chisso Corporation, a petrochemical company using a chemical called acetaldehyde, regularly released methylmercury into Minamata Bay. For years the residents of Minamata unknowingly consumed mercury-contaminated fish, resulting in their experiencing unexplained trembling, numbness, and blurred vision. The first case of Minamata Disease was officially recognized in 1956.

In addition to outwardly identifiable signs people exhibited, there was another more insidious invasion. Seemingly healthy young mothers were unknowingly poisoning their unborn babies with the mercury-laced fish they were consuming. Their children, born deformed and blind, were captured by the Smiths in photographs such as the now renowned image of *Tomoko Uemura in Her Bath*. Here, Eugene Smith uses the religious imagery of the *Pieta*. Draped across her mother's body, Tomoko is offered to the world. She suffers for all victims. Michaelangelo's *Pieta* uses scale shifts to impact how we perceive Christ in relation to Mary. The oversized scale of the Christ figure weighs heavily on the delicately rendered mother beneath. Where Knapp uses the scale of a doll to invoke an intimate relationship, Smith uses visual weight, composition, and religious reference to illustrate the martyred role this young woman and her family were thrust into.

It is Eugene Smith's impeccable formal compositions that draw us close to these stunning gelatin silver prints. They are exquisite in their composition. The richness of the blacks, their velutinous depths makes them seductive images. Their beauty compels us; we can't help but linger over them. Yet these images are painfully horrid. Susan Sontag tells us, "A photograph is supposed not to evoke but to show" (47). Smith's photographs do both. They show us the ravaged, deformed, afflicted bodies poisoned by mercury, and they evoke in us compassion and fear. Mercury's presence in Minamata Bay was invisible, yet the havoc it wreaked upon the inhabitants there is made glaringly visible in the photographs Eugene and Aileen Smith took. The invisible is the unknown, and the unknown can strike any one of us.

What compels us about these images is the human element. The people depicted are the unfortunate victims, other, yet they are people just like you and me. What makes us look at the images in the first place is that they are formally beautiful images. Therein lies the tension. The images are at times horrific, yet we are drawn to them. Their composition is astute and their execution pristine. Smith's gelatin silver prints are luscious. The blacks are like velvet and sit in contrast to the stretched and taut flesh of the hands around which the images are composed, a compositional element Smith returns to again and again throughout his career. With hands we eat, dress, bath, work, play, and connect one to another. An extended hand reaches out to welcome a new friend and is the first step in resolving a conflict. In a W. Eugene and Aileen Smith photograph of a Minamata victim's deformed hand, they are conveying to us the impact of mercury poisoning (see Figure 9.4). A hand so swollen that it can no longer function resonates in a way that a written report citing dangerous levels of mercury poisoning cannot. The photograph has immediacy; it connects the problem directly to us.

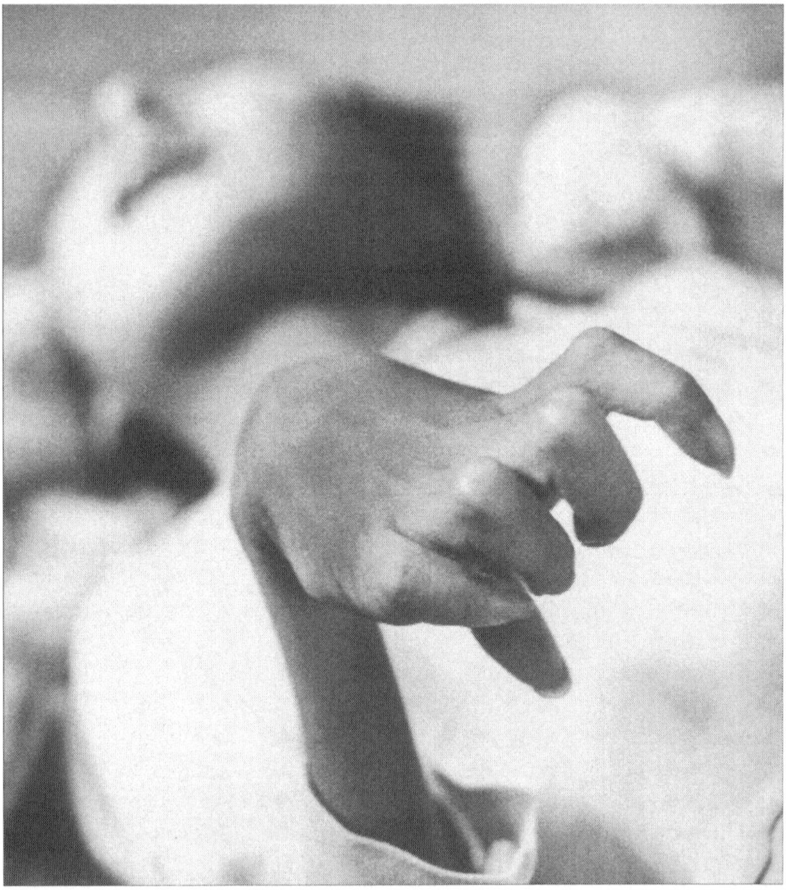

**FIGURE 9.4** *Tomoko's Hand.* Photograph by W. Eugene Smith and Aileen M. Smith, in *Minamata*, 1972.

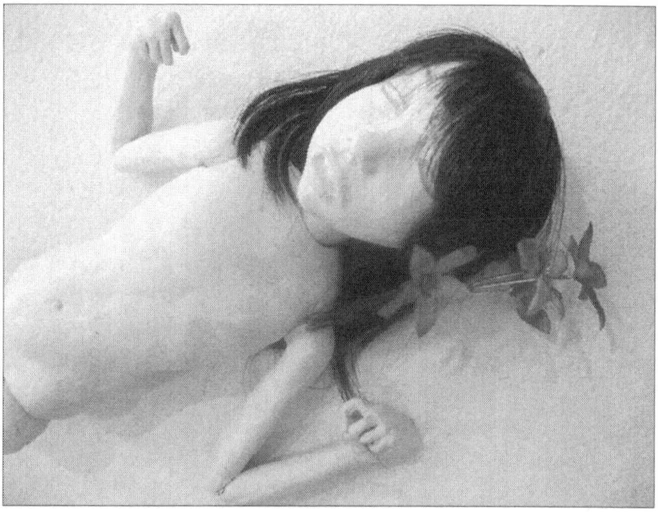

**FIGURE 9.5** *Kumiko* by John Knapp, 2008.

Another photograph by W. Eugene Smith of a child suffering the results of mercury poisoning inspired artist John Knapp's sculptural response. His sculpture *Kumiko* (2008) created in my "Sculpture in the Global Environment" class uses doll imagery to depict the crippled Minamata mercury victim (see Figure 9.5). "Wooden-doll" is a term frequently used to describe victims of mercury poisoning. Mercury poisoning affects a person's neurological system, and those afflicted lose control of their motor skills leaving them to hobble along as though on two wooden legs. The doll-like qualities create a tension between the innocence of a child's toy and that of a young girl unable to live a normal life—it shows her vulnerability yet evokes a presence that is tweaked into a peculiar place where the inanimate object becomes animate. The "doll" becomes a surrogate human and an almost adult at that. The scale is intimate, the size of a doll that a small child would cradle and carry around. There is a flower behind her ear, yet she is naked. Stripped bare. The flower represents nature and beauty. It is the hope we experience each spring as flowers bloom, promising a new season, new growth, and a new beginning. Yet her body lays twisted and deformed. Her hands, clenched fists, remind us of another Smith photograph, a close-up of Tomoko Uemura's hand, unable to hold even a simple utensil, rendering her completely reliant on those around her for an act as simple as eating.

Portland, Oregon, artist Linda Wysong's series of prints titled *Black Earth and Blue Sky* (2010) created in response to coal-fired power plants in the Portland, Oregon, area use a preindustrial art form, woodblock printing, to describe a postindustrial problem (see Figure 9.6). The grain of the wood present in the prints sits in distinct opposition to the mechanical power plant with its modern smokestacks dominating the image. The three smoke stacks remind us of Golgotha—or at the very least of a trinity; is this the altar of industry?

**FIGURE 9.6** *Black Earth and Blue Sky* by Linda Wysong, 2009.

An Oregon coal-fired power plant operated by a self-professed green company, boasting about its use of wind power among other things, sparked Wysong's attention. Issues of infrastructure have long been at the core of her artistic inquiries. Building, urban planning, sewage, solid waste, and the "guts" of the buildings we live and work in; places covered by the veneers of day-to-day life have long motivated her provocative installations and public works. She strives to expose the invisible. "The things we spend our money on show us where our priorities lie," says Wysong (interview). They shape our lives and our culture. The decisions we make for the built environment as compared to the natural systems found in nature tell us something about ourselves and about how we see our role in relation to nature and the environment.

What began as an investigation of a specific plant quickly expanded as Wysong realized airborne toxicants cannot be contained. We share air space. Pollutants coming from plants in China find their way to the western US coast. Suddenly, what was thought to be a local issue became international, lacking definable borders. Taking her clue from the sky and the clouds Wysong began the series *Black Earth and Blue Sky*. We all look at the clouds; we hunt for familiar images. We identify shapes as we laze about in the grass on a Sunday afternoon calling out, "look it's a rabbit; no, it's a butterfly." Using this friendly image she draws us in. For her 2010 installation of the same name she uses three large 5×6 ft. woodblock prints of clouds with a flash animation of the toxicants' representative symbols and the percentages of pollutants emitted from coal-fired power plants projected onto the prints, propelling an ancient process and the latest technology onto the same plane.

Wysong uses the graphic nature of woodblock printing to its fullest advantage. The bold marks demand our attention. They are not light or easily dismissed. These prints are steeped in the history of art. They recall the Works Progress Administration (WPA) murals of Diego Rivera and bold and provocative prints of Käthe Kollowitz. Wysong further pushes the history of art to her advantage with the addition of text on the bottom of the print listing the toxic emissions released by the plants. The inclusion of text recalls early protest posters of the Russian Constructivists. In the original, the text at the bottom of the print is in bright red ink (not reproduced here) and with coarseness of the letters read like a wound. The marks that compose the clouds are visually and symbolically laden, yet they whorl around bulbous and full like the clouds we gaze at as we lay about the park. They have a weight to them. The tool marks incise and gouge the wood the way Heizer gouged the earth in *Double Negative*. The cloud formations are of a scale that is human. When hung on the gallery wall they reach above us. The thick black ink impregnates the paper with the sensation of soot seeping into its surface just as the carbon monoxide will seep and gather in our lungs. The black-sooted carbon monoxide is made palpably present. It is the one thing we will see in this hideous group of contaminants. The mercury released into the atmosphere is invisible, made visible only in name at the bottom of the print.

The manner and tone in which environmental issues have been addressed by artists vary tremendously. From the depiction of the landscape in painting and drawing, to works made of the earth and about our relationship to the earth, art has been engaged in a dialogue about our ecological environment since humans first began making marks. One can argue that the first known images of animals recorded on the cave walls in Lascaux, France, in fact depict the landscape. Their placement on the cave walls creates a spatial relationship that implies landscape. This type of spatial arrangement is similarly found in Chinese landscape painting such as the earliest known painting of its kind, *Spring Excursion Zhan Ziqian of the Sui Dynasty* (581–618), depicting nobles playing in the mountains. While the cave paintings are corralling the objects, in this instance the bison they hunt for sustenance, the emptiness of the space surrounding the bison speaks to the vastness of the environment they inhabit. The land is plentiful and food in great quantity. While there is no direct image of humans, they are present in the hand that makes the drawing and in the eyes that are viewing; it is easy to see the balance they create with nature. The Chinese landscape painters share some similarity spatially but there is a stylization to the images that idealizes them. These are contemplative paintings conceived to nurture philosophical thoughts about beauty and one's relation to the land. Their intention is to bring the mind of letters back in tune with the sublime. Chinese paintings use isometric perspective. In this type of perspective, objects and figures, each the same size as the next, are placed spatially in the landscape. This kind of spatial proximity differs radically from our Western notion of representation where the illusion of space is mathematically devised. Freedom from mathematical restrictions allows the Chinese painter greater flexibility, they may choose to emphasize whatever they like through placement and size.

Western painting since the Renaissance has used geometric perspective to express space, reflecting the intellectual developments associated with that period. Here, the closer the figure to the foreground the larger it is rendered; the further away, the smaller. This perspective results in an image that attempts to record the landscape in the same way a camera does; it is a two-dimensional illusion of what we experience in front of us. In Chinese isometric landscape painting we are not lulled into the illusion of the landscape as our eye sees it before us; it is not about tricking the eye. Early Chinese landscape paintings tell us a story in a different way. They reflect a different relationship, one that has to do more with reverence and contemplation and less with depicting the exact idiosyncrasies of a given landscape.

Nineteenth-century painters like Fredrick Church and Caspar David Friedrich chose majestic views that reveled in the grandeur and spirit of the natural world. Their paintings evoke a sense of overwhelming awe and wonder, something greater than us. This differs wildly from the sparse economical compositions found in cave paintings and Chinese landscape painting. As we move around the world throughout history we find widely divergent ways of expressing the landscape in works of art. Each instant reflects the unique point of view and relationship a given population has to a particular place at a particular point in time.

Sculpture as well as painting fuels this relationship to the environment. Throughout art history artists have turned to the landscape as a way to understand where they are physically in the world. Depicting the landscape and interacting with it is another way to understand one's position in the world. In the '60s artists such as Michael Heizer and Robert Smithson created work using the land as their medium. Heizer's *Double Negative* (1969–70) was among the first "earthworks," artworks created as part of the movement known as "land art" or "earth art." *Double Negative* consists of two trenches cut into the eastern edge of the Mormon Mesa, northwest of Overton, Nevada. Rather than using the traditional approach of carving and constructing, *Double Negative* was made using heavy equipment to remove earth and rocks. In a sense it was a massive, large-scale carving, but rather than a carved stone or piece of wood the earth itself was the medium. The shift in scale from carving a discreet entity such as a rock or a stone to shaping the land itself, spanning some 1,500 feet, immediately places the viewer within the work of art, and in this case within the landscape, in a way that is external to our usual position. We become physically and mentally invested in the work of art. Suddenly the very world we inhabit is a continuation of the artwork, dictating our active role in it. The sculpture itself becomes the frame and we are inside the frame. The difference between this frame and the kind of frame in painting is that with the sculpture we are physically inside, whereas we psychologically enter the space of a painting by projecting ourselves inside the frame and into the world depicted.

Through the process of constructing Heizer's kind of art one is reminded of the act of incising used in ritual scarification. Here the act of cutting in, of opening up, functions similarly to an enlivening ritual. The same gesture that cuts also heals, swelling over and enlarging its physical presence. After the removal of several tons of rock and dirt, what is not there becomes as important as what is there.

This gesture is a radical one. The healing process is poignantly illustrated in sculptor and architect Maya Lin's *Vietnam Memorial* (1982) in Washington, DC.

Rather than a memorial erected high above our heads, here a v-shaped incision in the earth leads us to below the surface of the surrounding lawn. The path slowly takes us along a highly polished granite wall listing the name of each fallen soldier. As we descend along the walkway we are in effect walking into our own grave, looking at our own reflection in the polished granite as we read the list of names that cover the wall from top to bottom. Walking along the wall, accompanying the names of the fallen allows us the time to accept the loss and begin to heal. The crucial element here is the aspect of time as experiential. We move through space in time, the only way we can understand the landscape and our relationship to it is through the time necessary to traverse it.

Lin's *Vietnam Memorial* is not an aggressive act in the way that Heizer's work may be understood. The intention of the earthworks was to move art away from commodification. Some forty or more years later that notion is lost on a generation engaged in the preservation of the environment and a "leave no trace" attitude. My students today are interested in work that takes these attitudes into account. This is how they exist in the world and how they wish to see the world around them respond. A perfect example of this response is found in the more restrained approach of the work of the British artist Richard Long who uses the land in a very different way. He states, "I like the idea of using the land without possessing it" (Fuchs, 236). Long's approach is considerably gentler, lighter in its mark. Rather than use bulldozers and heavy equipment, Long walks.

> I, too, wanted to make nature the subject of my work, but in new ways. I started working outside using natural materials like grass and water, and this evolved into the idea of making a sculpture by walking. (Long)

Walking has allowed Long to expand the boundaries of sculpture to include the notion of place as well as its traditional roots in material and form.

> He makes a work of art in the place he chooses, with the materials that are there, on the slope of a mountain, high under the sky—a line of stones if there are stones lying about, or if the place wants a circle. When there are no stones, he can walk up and down the footpath as often as necessary for a line to become visible, or with his boot he can scratch a mark in the soft earth, or he can make a line by pouring water from his flask. (Fuchs 43)

Long's mark-making in the landscape reminds us of our continuous presence within it. Regardless of how subtle or slight that presence may be, we are reminded at all times we are having an impact on the land. The shift from the aggressive actions of the earthworks sculpture to the gentle impact we experience in the work of Long reflects a shift in approach and thinking that is indicative of a new dialogue with nature versus previous attempts to conquer or possess it.

In recent years many artists have been creating works that address issues of sustainability and ecological practices. Their work has helped to initiate a broader dialogue about the environment while at the same time continuing within a tradition of art-making that looks to a formal foundation of material and form. Formalism in art declares meaning in and of itself—by relying only on compositional elements and not acknowledging meaning beyond the material and form that composes each work of art. It is about the form over content. However, in the case of the artists cited here, context and meaning are layered over the formal framework allowing the pieces to resonate within a broader visual and cultural context. Existing outside the space of a gallery, the work finds an audience that extends beyond the usual art crowd. It becomes by virtue of its site more accessible to the lay crowd. The fact that the work is made in and of the natural environment is crucial to fully understanding its larger meaning. This meaning directs us to a greater engagement with the natural world and our role in maintaining a balance between nature and the world we build.

The artist James Turrell makes work that combines the approaches of Heizer and Long. In 1979, Turrell bought a 400,000-year-old, two-mile-wide crater. Roden Crater is an extinct volcano northeast of Flagstaff, Arizona, on the western edge of the Painted Desert. Here Turrell "carves" with light and space. He has moved tons of earth to carve viewing chambers and tunnels. Carved into the earth on a monumental scale, the chambers and tunnels form and frame the way we see and experience light.

> There's a space where you can see your shadow from the light of Venus alone—things like this. (I) also wanted to gather starlight that was from outside, light that's not only from outside the planetary system which would be from the sun or reflected off of the moon or a planet, but also to emanate light from the galactic planes where you've got this older light that's away from the light even of our galaxy. So that is light that would be at least three and a half billion years old. So you're gathering light that's older than our solar system. (PBS)

This approach radically changes the notion of time in a sculpture. Time is a fundamental element in sculpture. In order to experience a sculpture one must walk around it, move in space and in time. It takes time to understand three dimensions. Turrell's 400,000-year-old crater moves that notion of time to beyond the mere walking around and literally refers us back into the history of time.

It is the work of artists such as Patrick Dougherty and Andy Goldsworthy that my students today most respond to. Each of these artists works within a balance with nature. They make work of and within the landscape. Dougherty uses twigs and sticks to make his sculptures. While sturdy and durable in construction they are not meant to be permanent. Over time they will age and slowly deteriorate, returning from whence they came. Andy Goldsworthy's impact on the environment is extremely gentle as well. Goldsworthy's painstaking and ephemeral work uses only materials he finds on site often in the far depths of the forest. They, too, are temporary and often exist for the general public through documentation exhibited in galleries, museums, and books. Both artists continue a tradition begun by the land art movement that removes the preciousness of the object and its role as commodity and inserts it into part of the ongoing evolution of nature. In the end neither Dougherty nor Goldsworthy makes objects to be sold in the gallery. Their interactions with the land and its surrounding environments are recorded through photographs, and the actual sculptural elements left to dissolve back into the surroundings in which they find themselves.

Other artists deliberately work to create a dialogue with nature. They create interventions and interactions. For his sculpture *Host Analog* (1991) Buster Simpson took an old growth tree that had been felled in the Pacific Northwest and installed the cut-up sections in front of the Convention Center in Portland, Oregon. Building a system of water sprinklers around the sections, he brought the tree back to life as it sprouted and continued to produce new growth. Taking what was dead and resuscitating it, the work exists as a healing process. Simpson is offering us hope; *Host Analog* tells us we can make amends for our actions.

The history of art reveals to us the ways in which the environment has and continues to figure in works of art. The variety of ways this is achieved continues to develop exponentially, affording artists infinite ways of working with and within the environment. It has broadened the discourse concerning our interactions with the land and the vast issues that surround it.

Art is important in any discussion about the environment. The language of art—what makes art good—is the driving force behind the images that compel us. To achieve this force the formal and the conceptual are intricately linked. We can be drawn to work that is formal, "art for art's sake," and we are compelled by work that combines form and subject. But we are not compelled to look at subject-only-driven work. For a work to be compelling it must have form as well as subject. The Smiths' images of Minamata, Wysong's coal stacks, and Knapp's "doll" resonate because they walk so carefully between the world of form and the world of content.

Art serves as a bridge. It extends the built environment to the environs of the spiritual and the human. The desire to conquer and possess, I imagine, will never quite leave us. But through art, perhaps, we can learn to listen more carefully and to watch ourselves a bit more closely to find a synergy between the world we build and the one we inhabit.

## REFERENCES

Fuchs, R.H. 1986. *Richard Long*. London: Thames and Hudson.

Long, Richard. Richard Long Official Site. http://www.richardlong.org/.

PBS Online. *Spirituality, James Turrell. Art 21*. http://www.pbs.org/art21/artists/turrell/clip1.html.

Rawson, Joanna. 2009. *Unrest*. St. Paul, MN: Greywolf Press.

Reid, Herbert, and Betsy Taylor. 2003. John Dewey's aesthetic ecology of public intelligence and the grounding of civic environmentalism. *Ethics & the Environment*, 8(1):74–92.

Sontag, Susan. 2003. *Regarding the Pain of Others*. New York: Picador.

Wysong, Linda. Personal interview. March 11, 2010.

# 10 Writing as Environmental Stewardship

*Sydney Landon Plum*

## CONTENTS



> Those who contemplate the beauty of the earth find reserves of strength that will endure as long as life lasts.
>
> —**Rachel Carson,** *The Sense of Wonder*

## INTRODUCTION

One evening early in January 2005, I got stuck pulling off the road to visit our family's cottage in Warren, Maine. The cottage sits just above North Pond (which is really a lake—the nomenclature is a New England idiosyncrasy). The garage and the road, near which I was stuck, are on the slope above the cottage. The foot of snow left over from late December was thickly crusted with ice, and the tracks I had pulled into were slick. When the front of the car lost traction, the back was not entirely off the road. I have lived in New England for a long time; I should have known better than to pull into previously owned tracks in snow, as they are the first spots to ice over. I also did not ask myself who had made tracks in the snow in the field above my cottage. Standing outside the car, pretending to try to figure out how to get the car off the road, I saw that there were ice fishermen on the pond.

There are always ice fishermen on the pond in winter. This typical Maine pond is home to trout and pickerel, loons, mallards and black ducks, at least one heron and one kingfisher. Usually by late December the ice is thick. There are people who live around the pond year-round, for whom ice fishing is one aspect of a seasonally conditioned life. When I go to the town hall every summer to get a sticker for the dump (transfer station), there are handouts about both summer and winter fishing on the two ponds in Warren. The summer before I had been perusing the rack of pamphlets with information for fishermen. I had just attended a presentation explaining the chemistry of the transmutation of mercury as it falls out of the atmosphere and into the lakes and ponds of New England. Professor Susan Sutheimer of Green Mountain College patiently explained to a group that included few scientists how the chemistry of this cycle works and its impact on wildlife living in and dependent on this water. Her PowerPoint reminded me of what I love about science: the colored charts and swirling arrows—explaining how all life is interwoven. In the Warren Town Hall, next to the application form for a fishing license, was a pamphlet with the bold title: "Let's Get Mercury Out of Maine's Environment. You Can Help!" I took one from the rack and took it to the cottage to stick on the bulletin board where I keep the phone numbers for the transfer station and the video store.

While I was watching evening darken over the pond, standing next to my stranded VW Jetta, I saw two figures coming up the slope, pulling behind them a red wagon on runners. I used to pull my children up and down a dirt road in rural upstate New York in such a wagon, and for a moment I believed that the ice fishermen had a child with them. It was, after all, the Feast of the Epiphany and the stars were bright. When the men with their wagon were close enough to say "hello" I saw that they were pulling their augur. They were parked nearby—in fact, they admitted to having made the tracks I had pulled into. They seemed awkward about having parked near our garage, but I told them I had no problem with their using our property as access to the pond, as long they helped get my car off the road and the ice!

The next day I walked out onto the pond to chat with these same fishermen. I noticed them putting the trout they'd caught in their Coleman coolers and leaving the pickerel on the ice. I had already noticed, with wonder and admiration, the dozen bald eagles adorning the tree branches nearest the lake. The eagles were so confident in their right to those roosts that everyone simply walked under them. When they did take off it was an amazing feat of aerodynamics. They unfurled those six feet of wings and with one flap were far away. The fishermen explained that they left the pickerel for the eagles because they, too, admired their majesty. It didn't bother any of us that the eagle is a scavenger, like the crows by the roadside.

About one hour of research online confirmed what I feared. Pickerel, as a predator near the top of the lake's food chain, is full of the methylmercury that poisons every lake in Maine. The eagles, with sustainable populations just back from the damage done by DDT and a stint on the Endangered Species List, should not be eating pickerel. I pondered a nice way of telling the kindly fishermen to stop leaving pickerel on the ice. I decided to write a poem, in which I would compare the fishermen to the Wise Men wanting to give gifts to honor something miraculous, but suggesting that the gift they had chosen was inappropriate. I did write a poem, and gave it to a very few fishermen, but it was neither a successful poem nor successful proselytizing. I had more of an impact explaining the situation to a friend up the road who is also an ice fisherman. He smiled about the poem and passed the word to others on the pond. In the ensuing winters, I think fishermen truly have left fewer pickerel for the eagles, and I have seen the fishermen leave some trout. If the words I passed along had any part in changing the habits of the fishermen it is because everyone likes the eagles.

Why did I think that a poem would change the behavior of my neighbors? Many of us who live by the word also live with the notion that if we could just get the word out, people would read and—as if struck like Paul on the road to Damascus—would choose to live in a way that honored the lives of others before their own.

The literature written about the natural world may not have proven as effective as a blinding light, but for many readers it fulfills a deep need. Almost always there is a feeling of return, of the restoration of what may have been lost, when a piece of good writing takes us by the hand and leads us into the woods or the desert, down the mountainside, or along the riverbank. Just for a few moments we see a glimpse of the Eden that so many have imagined, or the childhoods we had—or might have had. And there are adventures. Not only do we get to experience, with particular illumination, places we may never visit firsthand, but sometimes we get to see them through the life of a turtle, a sanderling, an eel, or at the side of a lively otter or an affectionate lioness. We learn about the world and our humanity. We feel great sadness and deep joy. Often our souls are restored.

But are we changed? Does our relationship with the natural world change when we have experienced these sensuous, magical words? Do these magical words help us experience a richer kinship with the natural world more than, for instance, the pamphlet from the town office or a sharp editorial in the Sunday paper?

In my heart I believe that these are rhetorical questions, and the answer is "yes." In my teaching of environmental literature over the past decade I have been exploring the rhetorical possibilities of writers such as Rachel Carson, Edwin Way Teale, Gary Snyder, Linda Hogan, Terry Tempest Williams, and Henry David Thoreau. I have tried to find ways to have students consider not only whether reading these writers might change how they think and feel but also how they behave. What wonderful answers I have gotten—two girls from the city claimed to have spent the day picking up worms from the campus sidewalk ("Where they would get squooshed") and moving them to the grass. I have also gotten the dreadful answer: nothing changed.

## TEACHING POETRY TO AWAKEN ENVIRONMENTAL CONSCIOUSNESS

For the academic year 2009–10 I decided to take this inquiry—can literature effect change?—into my creative writing classes. During the fall semester I asked student writers to choose something they would like to see changed and create something that might reach others. The issues they chose included racism, health care, date rape, the destruction of coral reefs, and, with some nudging from me, the problem of methylmercury in the environment. For the spring semester I asked students to focus solely on the issue of methylmercury.

Writing about an environmental cause for any audience requires a unique blend of empathy, observation, knowledge, and aesthetics. The best activist artists represent relationships—between aspects of the natural world, human individuals, and human culture—in such a way as to draw listeners to the cause of sustainable and harmonious lives. In the short time afforded to this assignment, how would I introduce students to such a complex environmental issue and inspire them to invest something of themselves to create such a relationship? Furthermore, I needed to provide them with models of effective, activist art and, specifically, with a path to poetry.

The quintessential American activist writer of the past century is Rachel Carson, whose *Silent Spring*, published in 1962, has been judged by the New York and Boston public libraries as one of the most influential books of the twentieth century. For too long the success of *Silent Spring* was attributed to its timing; however, contemporary critics are beginning to delve into the craft that lies behind the impact of this book on the general public, the environmental movement, and policymakers.

Two recent collections bring considerable wisdom to bear analyzing and paying tribute to the literary talents of Rachel Carson: *Courage for the Earth: Writers, Scientists, and Activists Celebrate the Life and Writing of Rachel Carson*, edited by Peter Matthiessen, and *Rachel Carson: Legacy and Challenge*, edited by Lisa H. Sideris and Kathleen Dean Moore. An idea woven through many of the essays in *Courage for the Earth* is that of Carson's alchemical power to create poetic imagery out of scientific observations. As E.O. Wilson says, she synthesized disparate knowledge "into a single image that everyone, scientists and the general public alike, could easily understand" (Matthiessen, 2007b, 27).

The image of the silent spring created in the opening chapter of Carson's study of the dangers of overusing chemicals in the environment, specifically organochlorine pesticides, is "the brilliant metaphor that drew all these warnings to a point"—as Matthiessen notes in his "Introduction." In order to depict death Carson first creates a vibrant scene of life, with a past and future, with economic and social stability, with human and nonhuman life interwoven and interdependent. The death that comes as "a strange blight," an "evil spell," "a strange stillness" takes its victims in the specific ways that neurotoxins destroy life. Death is sudden and inexplicable, accompanied by violent trembling. Mammalian offspring are stunted, if live births occur at all. Trees do not flower. And there is stillness, this absence of life, everywhere: on the land, in the water, in the air.

Scientific accuracy embodied in a timeless, narrative form might be reason enough for the power that "A Fable for Tomorrow" wields. John Elder acknowledges these components in his contribution to *Courage for the Earth*: "much of Carson's remarkable power as a writer came from her mastery of imagery, her apt allusions to literature, mythology, and history, and the cadenced, musical quality of her language" (Matthiessen 2007a, 79–80). The rhetorical power of Carson's rhythmic prose writing is argued persuasively by Chiaki Asai in an essay on *Under the Sea-Wind*, in which Asai points out the techniques Carson borrows from other lyric poets writing of the sea—notably Swinburne and Masefield, for whom Carson professed admiration. Asai also discusses Carson's depictions of the nonhuman world: "For Carson, to get the feeling of what it is like to be a creature of the sea requires the active exercise of imagination and the temporary abandonment of many human concepts and human yardsticks" (Asai, 2007, 24). In this observation Asai is tacitly acknowledging that Carson's harmony with the Romantics was philosophical, as well as poetic.

Elder's review of the poetic qualities of Carson's writing encompasses not only her imagery and musical cadences, but her firm grounding in the Romantic traditions of poetry that are deeply critical of a mechanized view of nature. Like the Romantic poets, Carson's writing is informed by belief in the prophetic powers of poetry and hope nourished by close contact with the natural world. Elder identifies this tradition with the tradition of activist American nature writing:

> Nature writers and explorers like John Muir, activists like David Brower, and contemporary authors like Gary Snyder and Terry Tempest Williams all participate in a broad effort to assert the integrity and sanctity of the living world against an irreverent program of technology and profiteering. (Mattheissen, 2007a, 87)

While it should not come as a surprise to include Carson in the tradition of "unacknowledged legislators of the world," it might be a bit harder to remember, standing in that silent space, the river of hope passing through Carson's every work, including this dark one. Linda Lear makes reference to Paul Brooks' remark that *Silent Spring* was "a book about death, prompted by anger" (Matthiessen, 2007a, 22). Yet at the end of the fable the storyteller voices the need to explain how disasters such as those described might happen, manifesting another of Carson's strengths, another inheritance from Shelley, perhaps.

> Her love of the natural world allowed no separation between observed fact and the ethical responsibility to make that understanding part of public dialogue. Advocacy was inseparable from knowledge and understanding. (Matthiessen, 2007a, 20)

As the sixteenth-century courtier depicted by Sir Philip Sidney averred (in life and verse), poetry may be used to set the reader along a path toward love and a moral life.

Thus, to use Rachel Carson as an example of activist writing implies reinforcing my students' abilities as observers; focusing their imaginations on the creation of powerful images; reinforcing the music of our language; and sharpening their critical powers—while reminding them of the choices they make every day to participate in nature's unity or stand outside of nature (and hope and life). In one assignment! In about five weeks of instruction. It does sound impossible when stated this baldly. Furthermore, I was at a disadvantage that no teacher of nature writing should even begin to contemplate—I had no firsthand experience to offer them. It was not even possible to take them out on the ice of North Pond to view the eagles, meet the ice fishermen, and admire the pickerel laid out as an offering, or not.

On the other hand there was something intrinsically appropriate in working at such a remove from the sensory appeal of nature: methylmercury is a terrible, mysterious, and nearly invisible agent of death. Not until irreparable damage is done does the body betray. And even then rarely does either victim or observer understand what is at work beneath the surface. So another aspect of this writing experience might be to overcome the distance of abstraction to create something with heat and light.

"Those who contemplate the beauty of the earth find reserves of strength that will endure as long as life lasts." I have had this quotation from Rachel Carson in my daybook for years. I chose this quotation to introduce the assignment because I had decided that the seduction of writers (and readers) to this cause needed to begin in beauty. An awe-inspiring beauty. I wanted students to connect to something in the natural world that is threatened by methylmercury and to have that connection begin in the emotions.

Choosing loons was an easy decision, because nearly everyone who encounters loons is in awe of them. Birds that swim are mysterious. Large birds that swim and make sounds that make one's skin tingle are beyond mysterious. They are, perhaps, miraculous. The aloof quality of loons, evidenced by Thoreau's celebrated encounter, contributes to their mystery. And loons are seriously threatened by methylmercury.

My task was to create something that would introduce students to the majesty and mystery of loons. Unfortunately I do not have any photographs of loons, so I had to borrow them from Google images. (This means that, for the time being, the QuickTime movie I created can only be viewed by my students.) Still I was able to find excellent photographs of loons both above and beneath the surface of the water and to put together a five-minute show, in hopes of engaging the students.

I chose to try to seduce students into the watery world of loons before introducing the threat of methylmercury. The students viewed the images of loons for the first time accompanied by my descriptions of some of the more extraordinary habits of loons (using a circular provided by Maine Audubon) and a reading of David Wagoner's poem "Loons Mating," which describes their cry as "the beautiful sane laughter." The students heard various loon calls as they saw the images and, later, had time to freewrite. Then I ran through the movie again, this time explaining the effect of mercury on loons: smaller broods, failure to care for the young, uneven wing size resulting in an inability to fly. After the second freewrite I gave the students a handout about haiku and encouraged them to try to put their impressions from the two viewings into images and, if possible, into haiku.

In her introduction to *Haiku Mind*, Patricia Donegan explains this eponymous concept as:

> A simple yet profound way of seeing our everyday world and living our lives with the awareness of the moment expressed in haiku—and to therefore hopefully inspire others to live with more clarity, compassion, and peace. ... A fine haiku presents a crystalline moment of heightened awareness in simple imagery, traditionally using a kigo or season word from nature. It is this crystalline moment that is most appealing. However, this moment is more than a reflection of our day-to-day life—it is a deep reminder for us to pause and to be present to the details of the everyday. It is this way of being in the world with awakened open-hearted awareness—of being mindful of the ordinary moments of our lives—that I've come to call "haiku mind." (Donegan 2008, xi)

Haiku mind is a prerequisite for the poet of nature, who needs to see the world with clarity and understand it profoundly. Donegan's suggestion that the expressions of the haiku mind may also move others to a deeper understanding makes this practice particularly helpful to the activist writer:

> These uncertain times of our present world call us to go beyond the awareness of just the pear, and to become more conscious and engaged in our world—we can do this by using this awareness of haiku mind as a stepping stone and expand it. (Donegan, 2008, xv)

The connection between haiku mind and environmental mind is explored in Professor John Felstiner's interview with Gary Snyder, during which he asks, "Are haiku a concrete instance of our getting the environmental news from poetry?"

Another text looks at this connection through a study of traditional haiku:

> a number of geese
> migrating—in the sky
> not even a wake
>
> **Mori Sumio (b. 1919)**
>
> Haiku reading and writing can lead to a deeper understanding and appreciation for life, and in particular for our environment. When Mori Sumio writes of the few geese migrating overhead he recalls that in former days geese were much more plentiful, as

> reflected in both his own memory and the whole span of Japanese literature. Perhaps through his poem we will come to reflect upon the loss of geese in the modern world, and this may motivate us to 'live more lightly on the earth'—as some American Indians have expressed it.
>
> According to Bashō, putting our eyes and our thoughts on nature is the highest refinement of civilization. (Higginson, 1985, 258–59)

*The Haiku Handbook* is an excellent introduction to the form of haiku, providing a useful description of an English-language structure that creates the sense of rhythmical incompleteness of the traditional haiku. Higginson and Harter are quite firm on the necessary absence of metaphor or simile from haiku. The presentation of imagery clearly and concisely turns out to be a very difficult thing, although this example from the handbook, by Chiyo-ni, makes it look easier than it is:

> things plucked up
> all start to move
> low-tide beach

At the 2009 conference of the Association for the Study of Literature and Environment, Richard Kerridge, whose talks always illuminate the darkness, questioned whether the distance (physical and temporal) between ecological impact and where we live might be too great to be artistically represented. Here was another gap that needed to be bridged. I needed to construct an assignment based on the assumption that artistry can close these gaps. The haiku mind also recognizes the gap. "To protect and revere the imagination, we need to step beyond hope and fear and enter unknown territory. In Tibetan Buddhism this unknown is a *bardo*, or "gap," the word often used for the place between death and rebirth—yet it is also a place of in-between states of our so-called awake life" (Donegan, 2008, 150).

In class Tiffany volunteered her effort for class discussion:

> about to fly high
> wings spread out from body
> short, one long

Although a bit more explanatory than suggestive, Tiffany's haiku captures a piece of scientific information in the final image. During the second showing I read this: "Researchers found that loons with high mercury loads have unevenly sized flight feathers. Birds with wing asymmetries of more than 5 percent must expend 20 percent more energy than normal birds to fly, a deficiency that may impact their ability to migrate and maintain a breeding territory" (Biodiversity, 2008). The class was impressed that Tiffany had picked up this information from only one hearing and expressed the problem so succinctly in a clear image. (We did not discover that Tiffany is a science major until later.)

## MERCURY EPIPHANIES

Getting students to engage with loons at some level and the threats to them was only a beginning, as the problem of methylmercury in the environment is much larger than its impact on loons. After whatever "hook" is found to engage students with a writing assignment, I set them forth on a path of discovery. I needed to represent not only the breadth and depth of the methylmercury problem but also the means by which individual writers could enlarge their vision and understanding. I developed a PowerPoint presentation to show students three paths that they could follow to a creative witnessing of the threat posed by toxic mercury: the impact on nonhuman beings in the environment; the impact on humans; and the systemic quality of environmental toxins. I had information on each of these aspects and websites for students to visit, but I knew the futility of sticking it in a file for them to read without some incentive. And I needed to continue to make the connection between poetry and the awakening of their environmental consciousness.

I call the presentation "Mercury Epiphanies" as a reference to my own sudden awakening to this issue and to my own poem. Due to her generosity I was able to include portions of Dr. Sutheimer's presentation, with diagrams of the path of mercury through air into water as it transmutes into its toxic form. Other slides provide explicit information about the presence of methylmercury in wildlife: bald eagles, songbirds, polar bears, and alligators.

The PowerPoint includes a truncated version of the narrative of my moving from a general knowledge of the presence of methylmercury in the lakes of northern New England to writing a poem about the ice fishermen on North Pond. This introduction places side by side the pamphlet about methylmercury from the Maine Department of Environmental Protection and an excerpt from my poem about leaving the gift of pickerel for the eagles.

> The eagle and the evening turn
> fishermen to rapt magi and, pleased
> with what they've seen, they leave
> believing they'll redeem the gift betrayed.

In order to indicate that my choice to write a poem to change the habits of the fishermen was part of a poetic tradition, I quoted the Italian poet Salvatore Quasimodo who said in his Nobel Prize acceptance speech in 1959 that "The politician wants men to know how to die courageously; the poet wants men to live courageously." Rachel Galvin uses this quotation in her essay "Conversing with the World: The Poet in Society," implying that the poet rarely has opportunity to influence the business of contemporary American society. Still, in her essay Galvin mentions the infamous White House symposium on "Poetry and the American Voice" scheduled for February 2003, which was the impetus for the Poets against the War Movement. I don't think my poem influenced any ice fishermen, but at least it got me started on this pedagogy.

Now all I had to do was get the students moving along a path from general knowledge to local knowledge through a connection to imagery and then to poetry. I began where our knowledge of the terrible impact of methylmercury began—with the disastrous effects of mercury poisoning on the inhabitants of Minamata, Japan, in the 1950s. To tell students the bare bones of the story, I relied on excerpts from Douglas Allchin's eloquent narrative. However, for them to really begin to understand the impact of methylmercury they had to see Minamata. There is a scratchy, three-minute film on YouTube (it looks like the kind of newsreel we used to watch between features at movie theaters in the 1950s and '60s) that graphically represents the neurological effects of mercury poisoning on humans and on cats. I did warn the students before I showed the film that it was distressing, and I heard some intakes of breath during the showing. I complemented the film with some excerpts from writing about the illness from Oiwa Keibo's *Rowing the Eternal Sea: The Story of a Minamata Fisherman* (translated by Karen Colligan-Taylor, Rowman & Littlefield, 2001):

> I remember that it was around ten o'clock in the morning, on a warm day in September 1959. Father came walking toward us wearing only one straw sandal. I was sitting on a straw mat in the yard, and my mother and sisters were nearby. "I don't know why, but I don't feel well," Father muttered. (39)

Keibo's book tells the story not only of the early years of the disaster, when people were trying to figure out what was happening to this community, but of the community's resistance in the decades that followed and the attempts to get justice from the Chisso Corporation [see also Chapter 8].

Keibo's memoir is one creative response to the tragedy in Minamata. Another is the photography of W. Eugene Smith and Aileen M. Smith, most famously *Tomoko Uemura in Her Bath* taken in 1971 and displayed on the pages of *Life* magazine in an exposé that affected Minamata and the photojournalism that came afterward. Unfortunately, at the time that I was preparing this course I had not read Michiko Ishimure's eloquent *Paradise in the Sea of Sorrow*, a nonfiction collage of narratives, interviews, and firsthand impressions.

These texts and photographs are moving, but I wanted the students to have a multifaceted understanding of the dangers of methylmercury. Although these students probably have not had much experience of mercury thermometers, they are the right age to have been warned in high school science class of the danger of exposure to the liquid metal. Mattea Harvey's "Poem (To Be Read with Philip Glass's String Quartet No. 5)" depicts women doing industrial work with mercury. As with so many dangerous occupations, manufacturing products with mercury is traditionally the work of the poor, the disenfranchised, the West's "Others"—mercury thermometers are still made in India and China. Harvey's poem alludes to social justice issues, but focuses on the girls themselves and their fascination with the metal. This dark, associative poem is as spooky as Glass's music, which I played as background for a portion of the presentation.

> ... In the hot afternoon,
> the girls melt into various poses by
> the glass walls, molding their memories
> of the outside world into newer, clearer forms.
> One taps a finished porthole window
> with a small silver hammer and pronounces it
> sound. One runs her finger down the seam
> of a serving plate. Another holds a thermometer
> horizontally, and uses its markings to measure
> the height of the trees. The mercury inside
> shivers in the newly imagined breeze.

We had been engaged by loons, distressed by the damaged inhabitants of Minimata and the doomed cats, and mystified by young women playing with mercury, but to encourage activist writing we needed more of a local impact. I didn't want to simply illustrate the points about methylmercury the presentation had made, but to find images that would suggest connections to make and areas to explore. And, once again, I was hampered by not having relevant photographs of my own to use. I spent hours on the Internet looking for strong images that were free of copyright restrictions, and settled for some good images and two or three great ones (which are copyrighted). I feel that I relied too heavily on visual images, but I was depending on Pound's definition of the poetic image—"an intellectual and emotional complex in an instant of time."

The images I put together to try to convey intellectual and emotional complexities included images of a cement plant and its reflection on a body of water and a smokestack belching out faintly colored smoke; various representations of fishing; illustrations of birds and a photograph of carp in a pond. I really wanted to find a photograph of women working with mercury but could not. Students had time to freewrite while each image stayed on the screen, but I told them they did not have to write about each image—only the one or ones that really struck them.

Again, it is unfortunate that I had not seen the film *The Cove* at the time I put these materials together, as scenes from this documentary would be terribly effective in updating the presence of environmental and dietary mercury, as well as suggesting the tragic effects for the future.

At the end of this hourlong image fest, I told the students they were ready to explore one area of the issue of methylmercury—to use the images to engage curiosity and creativity. I had provided some assistance in the former pursuit by making folders containing information about different aspects of the methylmercury problem, some songs and poems that are tangentially related to the subject, and a few more images. Available to them were the story of the Andean native residents of Choropampa, Peru, who are dying of mercury poisoning; an essay by a woman suffering from mercury poisoning caused by her dental fillings; the lyrics to Mos Def's song "New World Water" and Marvin Gaye's "Mercy, Mercy Me"; Gary Snyder's poem "Mother Earth, Her Whales"; a photograph of crosses on a hillside in Libby, Montana, representing lives lost to industrial pollutants in that small community; an illustrated pamphlet created by the Indigenous Women's Mercury Investigation, under the auspices of the National Water Office, to spread the word about mercury, dioxins, and PCBs in the fish that these women eat regularly; and the Earthjustice Interactive Cement Kiln Map that shows the polluting cement foundry just about fifteen miles from North Pond, where I swim with loons. I gave a quick, guided tour to these files before I sent students off to pursue their interests and make their own discoveries. At the end of this preparation, I presented them with two quotations:

> The best emotions to write out of are anger and fear or dread. The least energizing emotion to write out of is admiration ... because the basic feeling that goes with admiration is a passive contemplative mood.
>
> **—Susan Sontag**

> Those who contemplate the beauty of the earth find reserves of strength that will endure as long as life lasts.
>
> **—Rachel Carson**

Their assignment was to collect more information, to note personal responses or stories, to find poems or cartoons to elaborate on their approach to the issue of methylmercury. I hoped that they would also begin to be aware of the emotions that would move them in discovery and writing.

I had to wait until the next class to find out if this experiment had worked at all, which images had sparked something in these students, and whether they wanted to write as a response to anger and fear or to preserve beauty and life. When they returned to class I encouraged them to talk, thinking a few might want to share impressions and ideas. I was bowled over when nearly every student in the two classes had something to say. The imagery had done its work.

In our discussion in response to the images, four of the first six students to speak were responding to the smokestack—and only one of them was a woman. I finally could not resist, but observed how very Freudian! The laughter probably made it possible for so many students to speak. Although many students were "aroused" by the smokestack, they had very different subjects in mind for poems. Nick Cinea framed this provocative question: Is building really an improvement? Alexandre Roy wondered: What can I make of the godlike image? Adam Penrod wondered: Do we understand that the smoke blocks out the sunlight? Jillian Wilkie thought of the environment defined by the smokestack and asked, "What about the people who have to both work there and breathe in the pollution?" In a similar vein, Christopher Kiertz tried to imagine what it would be like to work in the factory and see only the reflection of the smokestack on the water. For Jordan Kurtz the images of the smokestacks suggested this motto: "What's good for today may not be good for tomorrow." Jennifer Koenig saw the cloud of smoke above the smokestack and thought of illustrations of the home of the gods.

Other students responded to the photograph of the factory reflected on the water or to the distorted bodies of the Minamata residents. Students wanted to think about biomagnification and bioaccumulation; about the neurobiology of mercury poisoning (and how to represent it in images); about a woodblock print that represented fishing before industrial pollutants. They found their own images, and one discovered the relationship between mercury and the Mad Hatter of *Alice's Adventures in Wonderland*. Several were moved by Mattea Harvey's poem and the thought of young women, probably their own age, handling mercury and dreaming about it. Others were inspired by Carson's fable and hoped to emulate it.

As they worked on finding the metaphors and analogies in their notes, I looked at the notes I had just made and marveled again at the trust, curiosity, and ingenuity of students. In a few days I would be reading and responding to drafts of their poems.

At the same time the students were writing their poems they were contributing to an online forum about the assignment, which was part of the assignment. Also part of their assignment was a short narrative, due with the final version of their poems, discussing the possible ways their creative work might be used within their communities to promote a meaningful response to lessen the impact of mercury on the environment.

## REACHING A WIDER AUDIENCE

The final step in this assignment was for each student to think about a way in which his or her writing, or any piece of activist writing, might reach a wider audience in order to best effect change. The students were, to me surprisingly, quite thoughtful on this question. Lynnette spoke for all the students and all of the ideas when she made this observation:

> Poetry, for example, and haikus in particular, can put the pain of the moment or the joy of the moment straight into the hearts and minds of the reader without a lengthy introduction and wordy argument. The issue, however, is still with poetry's role in our lives today. We need to make poetry "cool" (to borrow from Colin) first, then we can expect to hit a chord with people. We writers know how to look at a poem and can feel the message, but what about the CEO of Dell, will he feel the poetry? He needs to be made aware that it's there first, that it's [sic] message is real first, and that it isn't just a loner in the corner of his book-filled room writing about how he "feels." That poetry represents the thoughts and concerns of many and that it still is a powerful way of conveying a message.

In our classroom discussions students suggested a plethora of ideas for dispersing poetry on the winds of change. It was widely agreed that our poems should be included in the Poetic Journeys program at the University of Connecticut. The creative writing program selects a few poems written by students each semester to send to the design center in the art department. These poems, or excerpts from the poems, are made into posters displayed on campus buses and the elevators in the library. One of the reasons I asked that students write short poems for this assignment was that the Poetic Journeys program prefers short poems. At the end of the semester each class will help select a poem from the activist writing assignment to submit to Poetic Journeys.

An academically oriented idea was that we "take the poems to the science buildings." It was unclear how this was going to happen, although one idea was to suggest that the writing courses offered in the science departments require reading and writing poetry. We do hope to put together a poster of our poems and ideas to send to academic conferences.

One student suggested we write our poetry as graffiti around campus, where students would stumble upon the poems. This evolved into more elaborate plans to create mosaics in the manner of the William Blake mosaics in London and distribute them where people would stumble upon them. We thought of other ways people might "stumble upon" poems. This discussion branched off into talk of pairing bits of poetry with images and getting them on T-shirts, notebooks, jewelry, caps, and many other places where they would attract attention. One suggestion was to submit a methylmercury poem T-shirt to the Spring Weekend T-shirt competition!

Several students already involved in spoken word poetry events felt we could adapt some of the poems for spoken word performance. As my student Joseph Campbell said, "I would go to different open mics and speak out on the topic a bit and then read the poem to show passion behind it." Joseph refers to this possibility in his poem: "You can feel this rhyme./But can we save us in time?"

Danielle Allen expanded on the idea of spoken word poetry, positing a Laurie Anderson-like performance, and more. "Spoken word, though, it would work fairly well, and would probably be where it would have the greatest impact. One could even do spoken word and incorporate a slideshow of photos or drawings, making it into almost performance art. I recently read a graphic novel about the destruction of the environment that had very poetic parts. I feel like this poem could use drawings or cartoons to emphasize the point of it and be more entertaining."

Alexandre Roy also saw a visual tie-in for his poem: "I feel as though my poem would be best utilized on a poster. While my poem is probably too long to be effective in a billboard type of setting, a stanza or two from it could be used in conjunction with the powerful image of the smokestack emitting its ominous black clouds."

Many felt that the poems could become songs and would be effective agents of change in this form. Jordan Kurtz envisioned his poem, "The Industrial Author," as a political song that would be available to fundraisers and activist groups to motivate communities to protect against the harmful effects of methylmercury. Robert Strauss wrote in his narrative:

> I read my current draft of my activism poem to my roommates. The three of us play in a band together and are always collaborating on new song structures and lyrics. I chose not to tell them that the poem was something I was writing for this class. They immediately responded by asking me if this poem was a song lyric written in abstract verse. Because of this, as I studied the poem more, I could see it right at home in the liner jacket of a CD or vinyl record. We played a short improvised jam using the verse from the poem as the lyrical content, and it actually sounded pretty good. The poem blended with the music in a manner that seemed surprisingly fitting.

Students worked on their poems through three or more revisions, and although there are a few examples of what I would call *tours de force*, none is yet perfected. The final product of the assignment is only part of what can be reviewed, with respect to its application to enlarging awareness of methylmercury toxicity. I constructed the discussion thread on our Blackboard Learning site and required the students to write a narrative in order to be able to review the process by which students learned about toxic mercury, became involved in the issue, and took that involvement into the world.

Looking back over the poems two things attracted my attention. During the writing and revising process I noticed how many poems were grounded on strong imagery. This might have seemed a given, considering their exposure to visual imagery in my introduction to the topic. Yet they had quite a bit of written material to work with also. I came to believe, based on reading the poems and the contributions to the forum, that it was the initiating use of haiku that brought forth these poetic images from the students. And that the poetry is much stronger for this initiation.

Starting these writers in haiku practice was serendipitous in making them focus intelligence and imagination on *bardo*—the gap that Donegan identifies as an aspect of the haiku mind and that I had hijacked from the realm of Eastern spiritual practice to that of Western ecological awareness. Poem after poem looked into the dark spaces left when species disappear from the environment; the dark skies created by industrial practice; the darkness of human sorrow; the dark space between the spiritual and the material worldviews. These young poets spoke of the darkness they experienced with sadness, irony, anger tinged with many other emotions, and sarcasm. Rarely are the voices in these poems defeated or cynical. They seem to me more often determined to begin to build the bridge over darkness by imagining the gap clearly.

Included in suggested readings as background for this assignment was an article that appeared in *The New York Times Sunday Magazine* on April 19, 2009: "Why Isn't the Brain Green?" by Jon Gertner. I included this article to impress upon students how difficult it might be to span the distance, described by Richard Kerridge, between ecological impact and where we live. I don't know if any students read this piece. Gertner's article—which talks about "frames" and "nudges," but not poems—is not hopeful. However, just as the semester and this assignment began, another article appeared in the magazine of January 31, 2010—"Is There an Ecological Unconscious?" by Daniel B. Smith. The examination of the new field of ecopsychology in this article intrigued me—and seemed much more hopeful in its argument than Gertner's—so I framed extracts from it for our online discussion: on Gregory Bateson's theory of ecology, Peter Kahn's research on rewilding, and Glen Albrecht's work on what he calls "solastalgia" and on the human sense of interconnectedness. Students responded to these with enthusiasm and insight; many of their comments are included in the appended material. As I read and reread Smith's article and the student responses, the question in my mind was: Is there a connection between writing about nature and awakening an environmental consciousness? Does the process of writing—perhaps, in particular, writing poetry—awaken whatever is necessary within the human to experience our original connection? I have no answer, of course, and I have not yet phrased this question for any of the psychologists and researchers mentioned in Smith's article. (This is one of my ongoing projects resulting from teaching this unit; the students have their own projects.)

However, I found these remarks by students about the writing process suggestive of the riches in this field still to be explored.

> Furthermore, the process of writing was rather enjoyable simply because I found the topic of mercury so engaging in itself. For me, it was a gateway to a larger discussion about quality of life, which allowed me to visualize its negativity in the physical form. In addition, because my prior familiarity with mercury was limited to that which resides in thermometers, I was able to create metaphors while enlightening myself at the same time by highlighting its seemingly harmless red appearance in thermometers yet acknowledging its truly silver and toxic form. (Kyle Hayes, revision narrative)

> The process of writing this poem was difficult from a scientific perspective. It is difficult to take a scientific topic and talk about it without using scientific terms. In science, one word is used to describe one aspect. One word conveys one thought, and one thought only. In poetry, multiple words can have the same meaning, and one word can have multiple meanings. Finding the right word to paint a certain picture is very difficult when words can have so many meanings in poetry. In addition, science is dry, matter of fact, and informational. Poetry can be information, but to have the right impact the musical component lacking in science must also be crafted. (Andrew Rice, revision narrative)

Andrew's observations oversimplify the distinction between scientific and poetic discourse, but he has discerned a pattern that is related to a distinction made by

Carson in *Silent Spring* and elsewhere. In *The Secular and Religious Sources of Rachel Carson's Sense of Wonder*, Lisa Sideris traces Carson's concern that the tight focus of scientific study be tempered by the humility and reverence that are a result of appreciating the mysterious universe of life.

> The specialist is the *bête noire* of *Silent Spring*. Numerous passages castigate the tunnel vision of specialists such as the chemical engineer peddling pesticides and urging the world to "beat its plowshares into spray guns."... "This is an era of specialists," she writes, "each of whom sees his own problem and is unaware of or intolerant of the larger frame into which it fits." (Sideris 2008, 244)

Kelly Shea developed the merits of poetic discourse in this contribution to our forum:

> I don't think it's so much reading literature of the natural world that makes us aware of our environment and our relationship with nature as it is writing this kind of literature. As an English Education major, I am a firm believer that truth and understanding occur because of a writer's interaction with what he or she is writing. I, myself, barely know what I am going to write until I start writing it. I have to tease it, edit it, work out the best way to translate thoughts into coherent sentences. Writing about nature forces the author to establish an intimate connection with it because the author must work very closely with his or her subject matter. Such an understanding of nature can be accomplished by reading about nature, but I think that writing about it is a more direct way to make a personal connection.
>
> For example, I read parts of *Walden* in high school and hated it. Then I read it again during my freshman year of college, and tolerated it a bit more. Last year, I read it yet again, and this year I'm rereading parts of it. It's taken a while (4 years), but now I feel as though I can appreciate what Thoreau is saying. Because of the nature literature I've read, I've even stopped listening to my iPod when I walk to class now in order to appreciate the sounds and sights of the natural world around me. This class is the first time I've ever really attempted nature writing, but I can already feel my relationship with nature changing. I feel a slightly stronger connection to the natural world around me, and it's only been a few weeks.
>
> The purpose of nature literature then, is to encourage the readers to write their own nature literature. Reading about nature should light a fire in the reader, and persuade them to start their own nature writing career in order to work on their own individual relationship with nature.

In his essay "A Long View of Rachel Carson," John Hay summarizes the gift that Carson cultivated, which was also the gift she left for us.

> No other naturalist has matched her capacity to bring together such well-honed skills of observation, such a profound understanding of interconnection within the natural world, and such a transcendent writing style. (Matthiessen, 2007a, 102)

This passage says to me, quite clearly, that if we want to raise a generation to eliminate toxic mercury from the environment, to roll back the impact of global climate instability, and to save the planet and all of its interconnected life, we have to teach them to write poetry.

## APPENDIX: SELECTED STUDENT POEMS AND COMMENTS

What follows is a selection of student poems showing the diversity of student responses to the imagery and information about methylmercury. Comments illustrate, to a limited extent, the dialogue that nurtured creativity. All student poems and comments are used with the permission of the student. A few students asked to remain anonymous, and their wishes have been honored.

"Match"
by Shanell Sharpe

They are companions
A match
Faces lit
Looking down smiling
Forgetting to look up
A tear falls from her red eye—in the pond
It ripples
Until it massively spreads
A massacre has started
Oh how he dreads
Dreads the fact that
The dry water reflects death
They are surrounded
They depart

Tiffany Schroeder, forum comment:

> I feel as though I am expressing the voice of those that are most in need; such as the fish, loons, and other animals. They do not have the ability to tell their story of how this chemical is harming and affecting their lives. It is our chance as humans to make the rest of the world aware of the methylmercury situation. Poetry is a helpful and creative way to spread the word of harmful effects of this chemical as well as what can be done. After writing this poem, I have an even more appreciation for the world we live in. Poetry for most is full of emotion. I think that a poem has more meaning than a standard editorial in a newspaper and would give readers an opportunity to connect on a more personal level with the topic. People must never forget that the earth was here before the human race and we should be aware of how we impact it.

"System Interrupted"
by Alexandra Sehulster

The Earth pays a deadly toll,
So we can burn our dirty coal,
Thinking we cannot hurt as men,
But then—

A cloud of smoke,
A child chokes,
Her body jerks.
The deadly liquid lurks

Disturbing the rippling lake.
Killing the fish, the birds, using life for its take.
And slowly the world will come to rest,
Leaving vacant waters, a deserted nest.

Alexandra Sehulster, forum comment:

Although there was not literally a diagram in the articles illustrating the process, I could see it anyway: all the animals/people represented by little pictures with big yellow arrows going from the power plant to the water, to the fish, etc. In my head, it looked like any other secure process learned in biology class, like photosynthesis or respiration. This process seemed as inevitable as any other natural biological process. This is what frightened me the most. ...

"The Water is heavy"
by Pat Notti

The Water is heavy today.
Mercury
Like a tumor
Its weight is measured in pain.

We hide from this truth.
Eyes shut
Breathe held
The Water is surrounding us.

We will not abandon this ignorance,
Until we are heavy ourselves.

Cortney Nadolney, forum comment:

If it negatively affects our wallets, we typically do not care to change what we are doing. Then with the Minamata pictures, I immediately thought of excuses people would use. Well, it is not happening to us so it is their problem, is one I thought would be prominent. This immediately led me to the poem where the writer discusses about how the Nazis came for others but he did not stand up for them because he was not part of that group. Then the Nazis came for him and no one was there to stand up for him. I used this form as a basis for my poem. But more importantly, I used this idea. We do not notice things that do not directly affect us. It is easier to pretend it is not happening than to get involved and try and change things.

"There is something in the water"
by Caitlin Avery Kroc

There is something in the water
In the fish
In our daughter

Deformed in anguish, all to nourish
Deranged, blind and bent
The girl: a broken doll
To God is her soul sent
Quiet, to appall, poison's enthrall

Photographs of endless pain
Death and lies
Minds that cannot contain
Agony and comfort vie, the small girl cries

No matter what you do
We are here with you

Emma Broadhurst, forum comment:

An image allows one to extract emotions, memories, thoughts, vocabulary, and connections from it. This photograph speaks for the devastation of the Minamata disease that ravaged so many individuals. This photo displays a mother–daughter relationship that many women and families can relate to, furthermore establishing that connection of the viewer to the message behind the image. This is the essence of activism, to have a creative response that is so stimulating to senses and feelings that are not comfortable yet we face them anyways because we become passionate towards the issue represented. ...

(Untitled)
by Jillian Wilkie

I am the destruction
that damages my backyard
I am the filth
that fills my community
I am the poison
that permeates my lake
And today, I do not change.

Jillian Wilkie, forum comment:

I can remember Earth Day, planting things, and learning how to recycle. These things have all been taught to me, but I do not maintain the routine of recycling and caring for the environment. Although I do not litter, dump chemicals, or anything that is outright damaging, I do not participate in anything that helps our environment either. ... (W)e must get "back in touch" with our environment.

"Dawn awakens the earth"
by Adam Penrod

Dawn awakens the earth,
Gives birth to life with sun's rays
Shining energy into world.

With life comes man.
Their lust for control drives them to build,
Spreading concrete over grass.

Covering green with grey,
Creating iron abominations
To harvest their desires.

Smoke blocks the light,
Shadow of structures darkens the day,
Lush colors turn to dull hues

We have constructed our tombs,
They stand all around us, looming.

Blocking the dawn, barricading life.

Forum comment:

I couldn't get over how majestic the smokestack looked. Though many of the other images in the slide show were interesting, the smokestack was really the only one that started my mind going, for lack of a better word. Like I described in my poem, the smokestack billowing the pure white steam hardly looked like it could be a representation of anything toxic or malicious. This is how the entire process or mercury bioaccumulation works, it is not evident at any stage in the process until it has reached toxic levels in the bodies of animals.

"Playground"
by Laura Titrud

Bravely embracing vertebrae,
she navigates through nerves
swimming up the spine,
leap-frogging large gaps
from synapse to synapse.
Cells slowly shrivel,
surrendering control.

Though she careens through blood streams,
tying up capillaries, capability
comes from outside—
To the flesh from fish
traveling in tides
by man's own hand.
A burial at sea, a quick fix.
A funeral hymn, but not hers.
Her solace will be found
in your cells.

Student revision narrative:

The difficult part was to put myself in the person's shoes. I had to feel like my life was being pulled away. I had no control over what was taking over my body. I felt gloomy, hopeless, depressed, and confused. I had to ask myself why is this happening? These were not easy things to imagine myself going through, but maybe that's the best way to understand things. Sometimes you have to go through things in life to live and learn.

"Fashion Week"
by Alexandra Sanders

Poisoned by the sea,
  the fish becomes
  part of me,
It's silver blood mixing
  with my narrow,
  red streams.
I'm dying to be thin.
  Invasive—like a parasite.
  It is here to stay—

I'm a plaything in society's
obsession with beauty.

The fish that sustains me
drains the life from me—

Wrapped in roe and rice,
my veins are cold
as ice—

I am dying to be thin.

Georgios Katsikis, revision narrative:

I instantly thought about my love for sushi and how I eat it despite some warnings of mercury. ... I thought that if I could relate this to myself I can do this to many other people to help them realize that this isn't some distant problem, this is something that happens every time we eat sushi sashimi or nigiri.

Bryan Banville, revision narrative:

My attempt was to show that Dr. Jekyll is the innocent bystander and that once he ingests the mercury he is overtaken completely by the negative effects transforming him into a different person who anguishes at the earth and natural world. The mercury overtakes him just like it has overtaken our greedy society preventing any change from happening.

"Mercury, an Epidemic"
by Jessica Thibodeau

Disfigured, Contorted
Reflexes fail as disaster strikes
The unsuspecting victims.
A slow progression—
Memories recede, waves at low tide
This time they will not return.

Twitching, Dying
The strained limbs of the prey
A mass of sharp angles.
A silent predator—
Metal as ruthless as it is bright,
Undetectable in the water's organisms.

Desperation, Madness
The inflicted succumb to manmade weapons
Held hostage in their withered bodies.
A swift pollution—
All around us lives are disintegrating,
Many unable to name their killer.

Jessica Mueller, forum comment:

Part of what is so disturbing about the whole mercury problem, is that it is not something that you can visibly detect. ... The other frightening thought is that this is not common knowledge, or human instinct to be cautious of how much fish we eat. A few hundred years ago when syphilis broke out, no one knew why everyone was in hysterics. Some people used the devil to explain the change in the community. The difference between that problem and this one is that we are too advanced in technology, resources and education today to not be informed about something like this. There is no reason why we shouldn't be broadcasting the issue worldwide, because it is a real problem, and there is a real solution. One of the most upsetting things about this disease that these victims suffered is that it could have been prevented, and now especially more than ever.

"Mercury Cycle"
by Kelly Shea
Bloated and sick with child I fled to the shores
to drop you from my womb among the pebbles.
I left you there, a mound of greying human flesh
Still coated in the slag of your birth.
Gnarled hands, strange lump of your spine,
Curious stub of a leg —
falling must have damaged you.
Mother of all, mother of denial,
mother of ignorance—I walk away.
But this water once bore multitudes,
and the shared burden of motherhood
calls me back. I lift you, caress you,
wade into the grey with a dangerous joy
even the poisoned can purify.
The tainted water stings like tears,
I cradle you beneath the waves,
hold you tight against my womb, try to put you back.
From the barren depths you gaze
With eyes that ask a single question.

Felicia Tiso, forum comment:

I think Mattea Harvey used women workers in her poem because they are the most at risk. They pass on the poison to their unborn children who are then affected from the day they are born until the day they die. They carry the children of the future, yet these children are born unable to care for themselves. The picture by Eugene Smith is hard to stomach. Because people chose to dump mercury into the water, that child as well as others will have to bear that burden for the rest of their lives. It is not fair nor is it just.

(Untitled)
by Brittany Warren

I do unto you as you do to me

You dislike how I destruct
Yet you do the same on me

You curse me for hurting your kind
Yet you do the same on me

You spread your poison like the population grows
I spread my poison like the wind blows

Only when my fury touches you
Do you revolt against me
Yet I do nothing to deserve this strife

But remember
I do unto you as you do to me

Eldar Kurtovic, forum comment:

Living in the city I did not really think about the healthiness of my community. In other words, if the city people decided to open up a power plant right next to our place, there was not much protesting that we would express. However in the suburbs everything was different. The way people respected the quiet area where we lived and the way they took care of the surrounding forest was remarkable. There was a proposal to open a Stop and Shop store in a small part of the forest. But people began writing petitions and getting signatures in order to prevent this. At the end the store was not built. What I am trying to say is that the people cared and did something about it; they protected their environment for whatever reasons they had.

(Untitled)
by Sonya Sinclair

Mother
Mother
Why must you deny me?
Mother
Mother
Why must you deny me?
The average test I cannot pass
Because you sat on your ass and denied me my rights to a proper nutrition

Mother
Mother
Why did you deny me?
Mother
Mother
Why did you deny me?
A little taste of fatty acids
And now my intelligence cannot surpass nor meet the normal.

Mother
Mother
Why didn't you fight it?
Mother
Mother
Why didn't you fight it?
The use of mercury
Which made links to poison my mind and now I die
As the fishes cry "Mercury Poisoning"
"Mercury Poisoning"
"Mercury Poisoning"
And the government's only concern is its economy
Its riches from its mines

Mother Nature, why?

Lynnette Repollet, forum comment:

Then again, it does seem to be man's ultimate desire to move forward.

"State Religion"
by Colin Neary

Worship It—
God of garbage that cannot be thrown away,
God of insatiable appetite—
Its concerns only wanton consumption, defecation.
All worshippers must be indicted,
Zealots feverishly murmuring prayers;
Poems written in Greed's pedestrian verse.

Powerless, responsible:
Serfs eeking out meager lives;
Meanwhile, maniacal malignant Mercury—
Agent of absolute terror,
Manufactured material of darkest nightmares
Rapes and pillages a planet
Only we can render uninhabitable.
Its promises:
Wealth, Luxury—Happiness.
Worship the phallic shrine
Even as it ejaculates into water—
Source of Life; impregnating Earth
Its cancerous seed of infertility.
Salvation awaits:
Savage, ravaged land,
Fog of acrid chemical fumes,
Lethal liquid that has been called water
As we enter a religious fervor—
Total, uncontrollable, spasmodic, writhing;
Body rotting inside-out ...
Promise of inexplicable Death.
Not heeding the Minamata warning:
Amnesia our cause of Death—
Death by Fish
As we try not to think about It.

Stephen Anstey, forum comment:

Perhaps one day I'll figure out a way to give the environment a helping hand but not until activists stop focusing on the big picture and start focusing on the little things we can actually do something about. In my opinion the secret to helping nature is to break it down, compartmentalize it into things small enough for any of us to help with and see the results of, people need a reason to do things and making them feel better about themselves by helping out the environment. ...

Forum comment:

I think the key to reaching people with ecological issues is putting them on a scale that they can comprehend and relate to. When we hear statistics about how many billions of

> tons of pollutants are being pumped into the environment, they are certainly shocking, but really it just makes the topic too unapproachable. ... What really got me interested in the mercury issue was when a friend of mine became ill from mercury poisoning from a high seafood diet last year.
>
> "There's a difference between safe and healthy"
> by Alexander Stackpole
>
> There's a difference between safe and healthy.
> There's a disconnect between us and our planet.
> There's a manner of disregard
> and a need to cut our losses.
> This by-product, this element of demise—
> how can it be so out-of-mind?
> When safety is synonymous with danger
> We need to check our status.
> When we all cheat—
> We all pay.
> Even those who never played.
> Or never had the chance to object.
> It doesn't matter how many decimals are left,
> or how many regulations are placed.
> The measurement might be small by itself—
> the summation of sources is anything but.
> There shouldn't
> be a need.
> It's all We've got.
> And it's all ours
> to let go.

Chelsea Lane, revision narrative and poem excerpt:

> For my second draft, I decided to do some research on how mad hatter syndrome came about, learning that the process of curing fur to felt caused this. I then wrote down all the symptoms associated with mad hatter syndrome, and used imagery and metaphors to express them. "A dusty, dimly lit factory/Flooded by a silvery poison."
>
> "Message from the Gods"
> by Jennifer Koenig
>
> Like Prometheus we stole fire.
> We hungered for more power,
> Erecting the smoke stack temple.
>
> We worship Mad Mercury.
> From Gaia to Neptune, He
> Spreads his message of murder.
>
> We devour toxic fish.
> A penny is all we need, as
> He delivers us to Charon.

Katie Townsend, forum comment:

> I believe the description of the gap is very accurate. Haikus are a representation of the in-between state when you awake and your mind really truly wanders. You are

able to think of brilliant relationships that you may not have been open to in complete conscious state. As I write more and more creative writing I witness myself falling into that "gap" more frequently and it opens me up to a whole new world of writing.

Robert Strauss, forum comment:

I really like this idea of "rewilding." With each generation we continue to get so wrapped up in new technology that we are becoming disenfranchised from nature. To restore harmony and balance to our ecosystems would be the logical first step in reinventing the commune with nature that we once had. People today seem to view nature as more of a novelty than anything else. The number of people who go on hikes or enjoy camping or who find inner peace and comfort simply by being in the wild is receding each year. Our relationship with nature reminds me of the stereotypical relationship that one has with an aunt that they aren't particularly fond of: it's nice to see once in a while, but it's usually just a pain in the rear and is something most people would be perfectly happy to do without. I read the Tao Te Ching a few years back and it completely changed the way I view and interact with nature. It talks of achieving an inner harmony and balance that is unattainable without surrounding oneself by the natural harmony and tranquility of the wild. Immersing oneself in the primal setting of the wild can be one of the most beneficial introspective activities I know of. If we were able to change our species' current relationship with nature we would be able to change our relationship with the earth, and therefore, with each other, for the better.

## REFERENCES

Allchin, Douglas. The poisoning of Minamata. http://www1.umn.edu/ships/ethics/minamata.htm (accessed January 25, 2010).

Asai, Chiaki. 2007. *Under the Sea Wind*: First in a treasure chest of sea books from Rachel Carson. Paper presented at the biennial conference of the Association for the Study of Literature and Environment, Wofford, South Carolina.

Biodiversity Research Institute. 2008. Mercury threatens the next generation of loons. http://www.briloon.org/resources/news-media.php (accessed May 16, 2009).

Carson, Rachel. 1962. *Silent Spring*. New York: Houghton Mifflin.

*The Cove*, DVD. 2009. Directed by Louie Psihoyos. Lionsgate.

Donegan, Patricia. 2008. *Haiku Mind*. Boston & London: Shambala.

Elder, John. 2007. Withered sedge and yellow wood: Poetry in *Silent Spring*. In *Courage for the Earth*, Peter Matthiessen, ed. Boston & New York: Houghton Mifflin.

Felstiner, John. 2007. The post natural world: An interview with Gary Snyder. The Poetry Foundation. http://www.poetryfoundation.org/journal/article.html?id=179396 (accessed September 27, 2009).

Galvin, Rachel. Conversing with the world: The poet in society. Poets.org. http://www.poets.org/viewmedia.php/prmMID/5904 (accessed September 27, 2009).

Gertner, Jon. 2009. Why isn't the brain green? *New York Times Magazine*. April 19.

Harvey, Mattea. Poem (To be read with Philip Glass's String Quartet No. 5). http://www.poetryfoundation.org/archive/poem.html?id=237184 (accessed September 20, 2009).

Hay, John. 2007. A long view of Rachel Carson. In *Courage for the Earth,* Peter Matthiessen, ed. Boston & New York: Houghton Mifflin.

Higginson, William J., and Penny Harter. 1989. *The Haiku Handbook*. Tokyo: Kodansha International.

Ishimure, Michiko. 1990. *Paradise in the Sea of Sorrows*. Translated by Lisa Monnet. Japan: Yamaguchi Publishing House. Originally published as *Kugai jōdo: Daini bu*. 1972.

Matthiessen, Peter, ed. 2007a. *Courage for the Earth*. Boston & New York: Houghton Mifflin.

Matthiessen, Peter, ed. 2007b. Introduction. In *Courage for the Earth.* 1–18, Boston & New York: Houghton Mifflin.

Muizainal. 2009. Mercury poisoning—the Minamata Disaster. 3 min. 11 sec. http://www.youtube.com/watch?v=ihFkyPv1jtU.

Oiwa, Keibo. Narrated by Ogata Masato. 2001. *Rowing the Eternal Sea: The Story of a Minamata Fisherman.* Translated by Karen Colligan-Taylor. Lanham, MD: Rowman & Littlefield.

Plum, Sydney Landon. 2005. "Eagles on North Pond." Poem presented at the biennial conference of the Association for the Study of Literature and Environment. Eugene, Oregon.

Sideris, Lisa H. 2008. The secular and religious sources of Rachel Carson's sense of wonder. In *Rachel Carson: Legacy and Challenge*, Lisa H. Sideris and Kathleen Dean Moore, eds. Albany: State University of New York Press.

Smith, Daniel B. 2010. Is there an ecological unconscious? *New York Times Magazine*. January 31.

Smith, W. Eugene. 1972. Minamata photographs. Masters of Photography. http://www.masters-of-photography.com/S/smith/smith_minamata.html.

Sutheimer, Susan B. 2004. Mercury: Implications for the northern forest. PowerPoint presented at the Nature and Culture in the Northern Forest Conference. Crawford Notch, New Hampshire.

# 11 The Necessity of International Agreement

*I.L. "Pep" Fuller and Clare Stankwitz*

## CONTENTS



> Mercury is a chemical of global concern due to its long-range atmospheric transport, its persistence in the environment once anthropogenically introduced, its ability to bioaccumulate in ecosystems and its significant negative effects on human health and the environment.
>
> **—Governing Council of the United Nations Environment Programme, 2009**

## INTRODUCTION

By the end of the 2009 meeting of the United Nations Environment Programme's Governing Council/Global Ministerial Environment Forum (UNEP GC/GMEF), the international community agreed to negotiate a treaty to control human-caused (anthropogenic) mercury pollution. Developing an international, legally binding treaty on mercury pollution had been considered repeatedly since the 1990s (van Hoogstraten, 2009). However, several nations, including the United States, had blocked these attempts in favor of voluntary programs and agreements—until the 2009 Governing Council meeting. In February 2009, for the first time since the UNEP GC/GMEF* had begun to discuss the matter, the United States announced that it now supported a new, free-standing, legally binding mercury convention (see Appendix II for the full statement of the United States at the UNEP Governing Council).

This announcement signaled a change in the US position on a global mercury convention that dated to the Clinton administration (van Hoogstraten, 2009). Decision 25/5 of the UNEP GC/GMEF thus set in motion the complex but necessary process of drafting "a global, legally binding international instrument" (hereinafter treaty) on mercury pollution by 2013 (Governing Council, 2009). Given the significance of the human-produced mercury contamination problems discussed in previous chapters, this chapter aims to elucidate the full significance of this treaty and the steps involved in its negotiation. Furthermore, since treaties are often necessary to address global environmental problems, this chapter also seeks to present the unfolding narrative of the treaty's negotiation as a relevant and informative case study for other issue areas for which global-scale environmental agreements might be developed.

First, the chapter will discuss why a solution to anthropogenic mercury pollution must be both global and legally binding. Then, model negotiation steps will be presented as a "recipe for a successful treaty," based on lessons from past negotiations of multilateral environmental agreements. Subsequently, several recommendations for international actions related to anthropogenic mercury will be presented. Finally, the steps necessary for the ratification and implementation of the treaty in progress will be discussed, with an emphasis on the procedures necessary for the United States to ratify the treaty. Each of these discussions will also examine general principles that can be applied to global environmental treaties more broadly.

## WHY A GLOBAL, LEGALLY BINDING TREATY?

Negotiating a successful international treaty is a multiyear process that requires much attention to detail and careful diplomacy on the part of the negotiators. It is anticipated that four full years will be required from agreeing to start negotiations to finalizing the language of the mercury treaty. This timeframe does not include the critical process of approval (ratification) of the treaty by the governments of a sufficient number of nations for the treaty to take effect, or "enter into force." Furthermore, when nations ratify a legally binding treaty they agree to abide by the stipulations of the international

* See Appendix I at the end of the chapter for a glossary of all acronyms used in this chapter.

agreement. Given all this effort, why is it significant—and necessary—to develop global and legally binding environmental treaties?

## The Case for a Global Solution

Many environmental problems, for example the international movement and deposition of pollutants or international trafficking in endangered species, are global in scope. Therefore, global solutions are necessary to resolving many environmental problems. This principle extends to mercury pollution as well. No matter the financial and technical resources available within a nation to control its domestic production of mercury, countries cannot effectively manage mercury pollution on a unilateral basis. Because it is an element, mercury cannot be destroyed. Thus, the only solution to mitigating mercury pollution is to control the production, use, and release of mercury that occurs through human activities such as coal burning, gold mining, waste incineration, and other industrial processes. Domestic regulation alone cannot control mercury pollution within a state's borders because of the quantity of mercury released into the atmosphere and hydrosphere globally, and the effectiveness with which it is transported for long distances and with which its metabolites (e.g., methylmercury) bioaccumulate in these systems.

Mercury released into the atmosphere from coal burning in China is deposited onto US soil and into US waters. Mercury released from gold mining activities in South America and chlor-alkali plants in the United States eventually drains to the oceans, where it accumulates to dangerous levels in larger fish such as tuna, consumed worldwide. Hence, mercury pollution does not respect national boundaries and is a problem that affects the global commons. With regards to the United States specifically, more than half of the anthropogenic mercury in US ecosystems is from international sources (Reifsnyder, 2009). This is often from mechanisms such as atmospheric deposition and the migration of wildlife that have accumulated mercury in their tissues. Consequently, the problem of mercury pollution cannot be resolved by isolated and uncoordinated actions of individual nations or by bilateral or regional cooperation alone.

## The Case for a Legally Binding Treaty

Solutions to environmental problems of a global scale must also be legally binding in order to be fully effective. Many nations have recognized the need for an international solution to mercury pollution and the limitations of acting only unilaterally in this area. However, reducing anthropogenic mercury pollution has long been stymied by a divide between nations that advocated for a legally binding treaty on the issue and nations that preferred voluntary partnerships to address global mercury emissions. Countries that supported only voluntary partnerships insisted that these partnerships were sufficient to address the problem, and preferred the flexibility of the commitments and actions necessary in a voluntary framework. They argued, among other things, that negotiating a treaty would require large amounts of time and financial resources, be less flexible than a voluntary mechanism, and be less cost effective because of greater administrative and institutional overhead required for a treaty (Governing Council,

2009). Not until February 2009, with the change in the US position, was this impasse resolved and the international negotiation of a legally binding treaty begun.

Mercury pollution must be addressed by a legally binding treaty for three key reasons. First, voluntary programs have been ineffective at reducing international, anthropogenic mercury pollution. Second, binding commitments are consistent in their stipulations and requirements despite regime change in the nations that are party to these commitments. Finally, despite the existence of several regional and global agreements on hazardous substances, none of these agreements address both mercury pollution and its global nature.

In 2005, an international program involving governments, intergovernmental organizations, and nongovernmental organizations (NGOs) was formed that has attempted to address mercury pollution and its effects through voluntary commitments of its members (UNEP, 2009b). This Global Mercury Partnership, administered by UNEP, was the favored method of addressing international mercury pollution for the United States and several other key nations until 2008, the end of the George W. Bush presidency. However, the voluntary programs at the core of the Global Mercury Partnership are insufficient for addressing this issue. So far the Global Mercury Partnership has been unable to control mercury pollution adequately, evinced by the recognition of the international community that a stricter method of pollution control is necessary in the case of mercury. Only eight nations and thirty-eight other entities are members of the Global Mercury Partnership (UNEP, 2009a), limiting the reach of its programs. Furthermore, such voluntary programs do not permit members to require certain practices or standards of other members of the program. Rather, the only commitments possible within such a framework are those that two or more members of the partnership decide to design and implement jointly. Without the structured commitments that would be agreed to in a treaty, nations do not need to implement mercury control measures that they do not desire to implement.

Voluntary systems' nonbinding nature and limited commitments predispose them to achieve results only marginally different than those that would be accomplished by uncoordinated domestic action alone. Complicating the possibilities for the success of the Global Mercury Partnership, verification of a member's claims about its practices involving mercury and mercury compounds is possible only on a limited basis. Partners in an effort undertaken within the Global Mercury Partnership have to rely on the word of other partners and the limited oversight of UNEP and the Partnership Advisory Group to determine whether other partners are fulfilling their commitments. Finally, enforcement of the commitments made under the Global Mercury Partnership is not possible, as no enforcement mechanism exists within the structure of the Global Mercury Partnership (UNEP, 2009b, 4). The overarching framework for the partnership also does not provide for any penalties for members who do not fulfill their commitments (UNEP, 2009b).

By contrast, many legally binding agreements do have enforcement mechanisms and specific ways to address noncompliance. One such example is the Montreal Protocol on Substances that Deplete the Ozone Layer, which has very explicit and specific procedures to be followed in the case of noncompliance with the treaty's stipulations by a party to the treaty (UNEP, 1998, 47–49). Therefore, a legally binding treaty on anthropogenic mercury pollution would allow for common commitments on pollution reduction, verification of the actions of parties to the treaty, and enforcement that is impossible under a voluntary program.

A legally binding treaty also provides consistency in the obligations of member parties, even when the regime change occurs in parties to the treaty. Treaties signed and ratified by a nation are binding on all subsequent governments of that state. Once a state has ratified a legally binding treaty and the treaty has entered into force, the state is bound by the stipulations of the treaty, regardless of the political attitudes of subsequent rulers or ruling parties. While a state could withdraw from a treaty it has ratified (van Hoogstraten, 2009), such action is rare. By contrast, in a voluntary partnership, partners in a program are essentially free to change or even retract their commitments. Such an action could occur if a government comes to power in one or more partner states that does not approve of the conditions of current partnerships. However, with a legally binding treaty all the parties to the treaty can be largely assured that the other states that have ratified the treaty will continue to honor their commitments. This lessens the threat to states controlling their mercury emissions, or other environmental problems in the case of other treaties, that they could be disadvantaged by other states failing to implement such measures. Thus, legally binding treaties on global environmental problems, including anthropogenic mercury pollution, provide global benefits. In the case of the developing mercury treaty, this will include greater and more consistent control of anthropogenic mercury, as well as the benefit for individual states of a more predictable international regime in this environmental issue area.

Given the advantages of an international legally binding treaty when compared with uncoordinated domestic action or even multilateral voluntary programs, it seems that such an agreement is the best option for controlling anthropogenic mercury pollution. Yet several international, legally binding treaties on various chemicals and heavy metals already are in force. Methylmercury, the most toxic by-product of mercury pollution, is an organic and persistent pollutant that bioaccumulates in ecosystems and is capable of being transported long distances. Therefore, it could potentially be considered for action under the Stockholm Convention on Persistent Organic Pollutants (POPs). Mercury is specifically included in the Aarhus Protocol on Heavy Metals to the UN Economic Commission for Europe's Convention on Long-Range Transboundary Air Pollution (UNECE LRTAP), which covers mercury, cadmium, and lead (UNECE, 2009) and has been ratified by fifty-one nations. Additionally, the Basel Convention on the Control of Transboundary Movements of Hazardous Waste and Their Disposal and the Rotterdam Convention on the Prior Informed Consent Procedure (PIC) both address aspects of the disposal and transport of mercury and mercury compounds.

Why then are these agreements insufficient to address the current problems of human-caused mercury pollution? In the case of the POPs agreement methylmercury is not currently included in any of the three annexes of chemicals controlled under the POPs agreement, and thus cannot be controlled internationally under this agreement (Stockholm Convention, 2008c). It could be argued that methylmercury should be covered by the Stockholm Convention. The convention does have an adding mechanism, known as the POP Review Committee, by which methylmercury could become a chemical controlled by the POPs agreement (Stockholm

Convention, 2008a). However, methylmercury has not been submitted to the POP Review Committee for inclusion in the lists of chemicals covered by the POPs Convention (Stockholm Convention, 2008b). This exclusion is in part because during the negotiation of the convention, the United States and others argued that methylmercury should not be included since mercury is inorganic, and therefore the unintentional production of methylmercury through the alteration of mercury in biogeochemical processes should not be controlled by an agreement focused on organic pollutants. In the case of the Aarhus Protocol the LRTAP convention on which it is based is only a regional agreement, housed in the UN Economic Commission for Europe (UNECE, 2008a). Therefore, while multilateral, this protocol is not global and thus excludes major emitters of mercury such as China and other countries where coal burning and artisanal, small-scale gold mining is common. In addition, it is restricted to air pollution and thus does not deal with other pathways of mercury contamination including pollution of lakes, rivers and oceans, or bioaccumulation in migratory species.

With regards to the Basel and Rotterdam conventions mentioned earlier, the preparations for the mercury Open-Ended Working Group considered the possible intersections between the work of the mercury convention and these two conventions (UNEP, 2009c). As discussed in a preparation document for the Open-Ended Working Group, the Basel Convention is developing guidelines related to the management of mercury wastes. The negotiation of the mercury convention aims to not duplicate this work, but the Basel Convention does not deal with the anthropogenic production of mercury or mercuric compounds (UNEP, 2009c). The Rotterdam Convention addresses international trade procedures for several mercuric compounds, but again does not cover anthropogenic emissions or productions of elemental mercury or organic mercury compounds. Therefore, neither elemental mercury nor its inorganic or organic compounds are fully addressed in current multilateral, globally binding environmental agreements. Hence, the new convention on anthropogenic mercury pollution is imperative, and it is critical that negotiations proceed as smoothly and successfully as possible. The following section outlines key negotiation steps and puts forward recommendations for the provisions of the new treaty.

## RECIPE FOR A SUCCESSFUL TREATY

Negotiations of the multitude of existing global environmental agreements have established a largely standardized process for the development of new, international legally binding treaties. These previous negotiations hold many lessons for the negotiators who will be responsible for the creation of the mercury treaty, as well as other multilateral environmental agreements to come.

### Outline of the Negotiation Process

One of the main considerations of the negotiation process is the extended length of time required for international negotiations. In the case of the mercury treaty, developing an agreed treaty text will require four full years comprised of at least five meetings lasting around a week to two weeks each plus intersessional work between these meetings. To begin this process the international community must receive a strong mandate for the commencement of negotiations. With the mercury treaty, this happened in February 2009 at the 25th UNEP GC/GMEF. In the fifth decision of this meeting, the Governing Council agreed to develop a "new global legally binding instrument" on anthropogenic mercury pollution "which could include both binding and voluntary approaches, together with interim activities, to reduce risks to human health and the environment" (Governing Council, 2009, 21). Through this decision the Governing Council gave the international community and the executive director of UNEP clear instructions to begin negotiations, along with a list of specific provisions to be included in the treaty (Governing Council, 2009, 21–23). These instructions include continuing current work on mercury pollution and convening an Open-Ended Working Group (OEWG) and an Intergovernmental Negotiating Committee (INC). Such a mandate is necessary to begin the negotiation of any multilateral environmental agreement.

Convening the OEWG is the first preparatory step in the negotiation process after this mandate has been established. While the OEWG has only one meeting, it is an especially important meeting to the negotiating process because the OEWG recommends the agenda for the INC and describes its parameters in more detail than the mandate. Therefore, the OEWG ideally addresses procedural and some substantive matters in advance of the INC meetings so that negotiators at the INC can concentrate more thoroughly on the actual negotiation of the treaty. Between the OEWG and the first meeting of the INC, the secretariat continues to develop documents for the various INC meetings (Ashton, 2009). Hence, the next step after the OEWG and follow-up work by the secretariat is the meetings of the INC.

Five INC meetings are currently planned for the mercury treaty (Ashton, 2009). The INC has been mandated to have the treaty text ready before the 27th Governing Council meeting (Governing Council, 2009, 21), which will occur one month after the 2013 (final) meeting of the INC. After the negotiation process is completed, the treaty will then be opened for signature and ratification. Most nations that have participated in negotiations sign the treaty at a ministerial level meeting after the final text has been translated into the official UN languages. Some nations may also ratify the treaty at the final INC meeting. Other nations, such as the United States, require the time needed for a domestic process prior to ratification. Once a sufficient number of countries have ratified the treaty, a number usually specified in the text of the agreement itself, the treaty enters into force. Only after the completion of this step will the treaty be legally binding on countries that are party to it.

## The Open-Ended Working Group and the Negotiation Process

Ideally, in any negotiation, the OWEG should accomplish the tasks necessary in time for the substantive work of the treaty to begin at the first meeting of the INC. These tasks include recommending some sources of scientific input for the treaty and the structure of the INC itself, including the timing of INC meetings and who should be on the bureau of the INC. For the mercury treaty, the OEWG met from October 19–23, 2009, and discussed several topics related to the INC. It "recommend(ed) rules of procedure to the INC, as well as intersessional work for the Secretariat. ...

Delegates also took part in information sessions on supply and storage of mercury, artisanal and small-scale gold mining, and products and process" (Ashton, 2009). Thus, with regards to the organization of the INC, the working group developed proposed rules of procedure and a timetable for the negotiations, bearing in mind the timetables for negotiations occurring in other forums on similar issues (Ashton, 2009). Sweden updated the OEWG on its preparations for hosting the first INC for the mercury treaty from June 7–11, 2010. In relation to the substantive work of the INC, the OEWG covered available information on artisanal and small-scale gold mining, as well as the production, supply, and storage of mercury and related wastes (Ashton, 2009). The OEWG also received an update from the secretariat on the structure and timeline of the "study on various types of mercury emitting sources" that was requested in Paragraph 29 of the original Governing Council mandate (Ashton, 2009, 3). Finally, the OEWG received nominations from most regions on the countries that will comprise the bureau, or central governing body, of the INC. The Czech Republic, Jamaica, Mali, Nigeria, Russia, Sweden, the United States, and Uruguay were nominated as bureau members.

As described above, the OEWG for the mercury treaty has identified some of the scientific underpinning for the discussions of the INC. The commissioning of the "Paragraph 29" study by the Governing Council has somewhat simplified this process, and such a step should perhaps be considered in the development of future environmental agreements. While scientific input from other sources will undoubtedly be needed and used, this study will serve as a major source of information for the INC, and it is anticipated that it will be ready by the second meeting of the INC (UNEP, 2010). The Paragraph 29 study will cover, among other scientific and technical matters, "various types of mercury-emitting sources, as well as current and future trends of mercury emissions, with a view to analyzing and assessing the costs and the effectiveness of alternative control technologies and measures" (Governing Council, 2009, 22). At the OEWG, it was clarified that the Paragraph 29 study "will be focused on four sectors: coal fire power plants and industrial boilers; industrial metal production; waste incineration; and cement factories" (Ashton, 2009). Additional requests were also taken by the chair of the OEWG for additions to the study regarding issues that were not necessarily included in the original mandate for the study (Ashton, 2009). The OEWG consequently laid much of the scientific groundwork for the INC, clarifying sources from which the INC can obtain the scientific information necessary for its decision-making processes (Ashton, 2009). Testimony was also heard at the OEWG from several interest groups that will provide scientific and other input into the INC, such as the Artisanal Gold Council, the Mercury Policy Project/Zero Mercury Working Group, and the Natural Resources Defense Council (Ashton, 2009). Such interventions can be quite powerful. In the past the interventions of indigenous peoples who were affected by global circulation of persistent organic pollutants (POPs), but lived in regions where POPs were neither produced nor used, had a profound effect on the countries negotiating the Stockholm Convention on Persistent Organic Pollutants.

Regarding the structure and timing of the INC meetings, the OEWG clarified several points. First, the OEWG debated the rules of procedure as set forth in one of the preparatory documents for the working group meeting. This debate concluded in the rules being recommended to the INC for adoption, as they were drafted for the third preparatory document for the OEWG (Ashton, 2009, 1–2). With regards to the structure of the treaty itself, the OEWG conducted an open debate and decision-making process on how to develop this structure. The working group recommended to the INC what types of decisions should be left to the countries of the INC or left to the secretariat during the negotiation process (Ashton, 2009). In a further reduction of the work left to the INC, the OEWG also came to a conclusion on the timetable of the negotiations, agreeing on the timetable set forth by the secretariat in the relevant preparatory document for the working group. However, the OEWG did not reach a conclusion on the full membership of the bureau of the INC. The working group received nominations from all regions except the Asia-Pacific region. Therefore, nominations for this final bureau member will have to be received by the INC. During negotiations a well-functioning bureau can assist in obtaining consensus in plenary meetings and in avoiding spending excessive time debating matters in plenary that are critical to moving the process forward but do not require the deliberation of all nations participating in the negotiations.

### The Intergovernmental Negotiating Committee

While in the case of the mercury treaty it took more than a year of preparatory work to reach the first meeting of the INC, it is in the INC meetings that negotiation of treaties begins. Negotiations are based on the mandate for the treaty and the recommendations of the OEWG. Thus, the principal role of INCs is to draft and finalize the actual provisions of treaties. In accomplishing this INCs cannot be inconsistent with the mandate for the treaty, given by the Governing Council in the case of the mercury treaty. The INC for the mercury treaty has three main objectives when completing this task. First, it must develop a treaty that effectively addresses anthropogenic mercury pollution. Second, the treaty must be drafted to gain the support of enough states that it can be ratified by a sufficient number of countries to enter into force. Finally, states that are key contributors to the problem of anthropogenic mercury pollution, such as the United States and China, must be willing to ratify the treaty that results from the work of the mercury treaty INC. In the first mercury INC positions on a number of issues were put forth from many delegations. A wide range of topics were discussed that related to treaty objectives, structure, compliance, technical assistance, mercury trade and wastes, atmospheric emissions, raising awareness, and scientific issues. The secretariat also presented a draft of final provisions for the treaty at the end of the session. However, most decisions were deferred to the second (and later) meetings of the INC.

Clearly, the task set forth for INCs is challenging, even daunting. With the lack of formal organization between countries, especially when compared with most national legislative processes, how can nations meet these challenges when similar domestic legislation might be difficult to achieve? One key lesson from previous negotiations is that the work of INCs cannot be completed only within predetermined meetings. Though there will be five meetings of this body for the mercury treaty, and these meetings will at times involve slow and laborious discussions, countries cannot expect all the necessary agreements and work to occur in these sessions.

Hence, the mercury treaty INCs, and other future INCs, will likely be successful only if countries make an effort to meet outside of scheduled meetings. Such outside work is frequently bilateral (between only two states) and should be focused on mutual understanding of the objectives of each party as well as identifying areas of mutual interest. In addition to intersessional meetings, it is important to allow time during the formal INC meetings for informal discussion among government representatives, particularly when the plenary may have reached an impasse. The establishment of expert groups, once major objectives and concerns have been addressed, is one method of promoting these discussions. In this way, consensus on the treaty provisions can be expedited. The likelihood of leaving out a major concern of a potential party to a treaty is also minimized, increasing the chances that the treaty can rapidly enter into force.

## Comparison of this Negotiating Process to the Processes for Other Treaties

The negotiation process for the mercury treaty, as discussed above, is overall quite similar to the negotiation processes followed for other treaties. As expected, the steps of first obtaining a mandate, then proceeding to a working group, a series of INC meetings, and the procedure for entry into force are modeled directly on previous processes.

However, the planned stipulations of the mercury treaty differ in one significant way from previous chemicals treaties. Agreements such as the Convention on Long-Range Transboundary Air Pollution, the Basel Convention, the Stockholm Convention, and the Rotterdam Convention cover many pollutants that are chemically similar, or cause health and environmental problems through similar means. With the LRTAP, all the chemicals covered by this protocol were transmitted internationally through atmospheric processes. With POPs, all the chemicals this agreement covers are organic (carbon-based) compounds that bioaccumulate and remain in the environment for long periods of time without breaking down into other, less harmful compounds. The Basel and Rotterdam conventions cover the problems generated by the international transport and disposal of hazardous chemicals.

Avoiding an unnecessarily narrow scope is critical in any environmental treaty, as demonstrated by these examples. The Stockholm and Rotterdam conventions are treaties that name a small number of high-risk chemicals that have been banned or severely restricted, but both treaties have provisions for adding to the list of chemicals they regulate in future. Hence, these treaties are "ambulatory," capable of growing to include chemicals that may in the future fit the criteria for regulation under these treaties. Environmental agreements should be structured with a broad or expandable scope for regulation due to the time, energy, and costs associated with the negotiation of a legally binding international agreement. By contrast, the new mercury agreement is currently set to cover only one element and its compounds, despite discussions prior to the treaty mandate about controlling anthropogenic lead and cadmium through the upcoming agreement. Both of these substances are, like mercury, heavy metals present in elevated, toxic levels in many ecosystems due to human activities that use and concentrate these substances. Though the mandate for the treaty does not prohibit the inclusion of additional heavy metals in the treaty, several key nations, including the United States, have indicated that they only want to address mercury in this treaty.

This narrow approach appears to be a vestigial remnant of the previous administration's policies and overall negative position on a global, legally binding mercury treaty. As explored previously, it is certainly necessary to control anthropogenic mercury. However, this single-element approach risks exacerbating fragmentation in the international chemicals control regime and costing the international community more time, energy, and funds to regulate other heavy metals at a later time. By including twelve different persistent organic pollutants in the POPs agreement and creating a scientifically based adding mechanism, the international community did not have to negotiate a new treaty for each pollutant. Yet, with the way that the mercury treaty is planned, lead, cadmium, and other heavy metals could not be regulated without the negotiation of new global agreements. Moreover, the inclusion of an adding mechanism in the Stockholm Convention allowed for POPs identified after the negotiation of the treaty to be managed under the same agreement, by the same secretariat and staff. Thus, an adding mechanism in the mercury treaty would simplify the work of the international community, should states choose to regulate other heavy metals on an international basis in the future.

A narrow regulatory scope for a treaty presents not only possible difficulties for the international community in the future but also problems for domestic regulators and technical support staff who will be responsible for the implementation of a treaty and its provisions. At least in the United States, the regulators who will be in charge of implementing treaty provisions on a national level do not deal with issues as narrow as a single substance. Rather, environmental regulatory agencies have staff that are specialized in and trained to regulate many related substances or other environmental issues. In the case of mercury domestic regulators in the United States work with heavy metals and not just mercury. In addition to this domestic component specific requirements of the mercury treaty will involve developed countries providing technical and other assistance to developing countries, as is the case with many previously negotiated multilateral environmental agreements. According to Decision 25/5 of the UNEP Governing Council (Governing Council, 2009, 22), the mandate for the mercury treaty, "the ability of developing countries and countries with economies in transition to implement some legal obligations effectively under a legally binding instrument is dependent on the availability of capacity-building and technical and adequate financial assistance." Thus, developed countries will be sending technical support, including staff, to countries that require such assistance.

Again, those providing this assistance frequently are involved in regulation of other heavy metals as well and could give valuable advice and assistance on a number of heavy metals that pose cross-boundary health and environmental threats. The lack of inclusion of other heavy metals in the mercury treaty thus also complicates the provision of such assistance. Therefore, a key recommendation of this chapter is that the United States change its opposition to the inclusion of an adding mechanism for other heavy metals such as cadmium and lead. Given the challenges faced by developing countries in meeting many of the requirements of multilateral environmental agreements and the continued commitment of UNEP to sustainable development, such assistance will likely be a stipulation of any future environmental agreement. Therefore, this reason for broadening the scope of the mercury treaty applies to the future negotiation of other multilateral environmental agreements.

## US Negotiators in the Treaty Process

As mentioned, the change in the US's position has been critical in moving the international community from voluntary actions on anthropogenic mercury pollution to developing a global, legally binding treaty on the issue. The United States has also played critical roles in the negotiation and effectiveness of other treaties. US leadership and early action on ozone-depleting substances was critical in the success of the Montreal Protocol on Substances that Deplete the Ozone Layer (Smith, 2009, 151). US obstruction of negotiations has also severely hampered the efficiency and effectiveness of several treaties. As a result it is important to understand the role of US negotiators in the treaty process, and how the US delegation can contribute to the success of the negotiations and the treaty. First, it must be recognized that the primary role of all delegates to an international negotiation is to convey and advocate the positions of the governments they represent. In the United States, these positions are developed in prenegotiation discussions among the relevant federal agencies, with the Department of State frequently acting as the party that is relatively neutral as far as domestic interests are concerned (Smith, 2009, 146).* On its surface this may lead to an impression that producing a treaty where more than one party is satisfied by the outcome is impossible. However, the ability of governments to reach the level of agreement necessary for a treaty is greatly dependent on how the objectives of the governments are defined.

In addition, compromises are frequently employed to satisfy multiple parties with dissimilar but noncompetitive objectives. Under the previous two US administrations, the objective of the US government, and thus of US negotiators, was to avoid binding international obligations to reduce anthropogenic mercury pollution. These objections of the United States to international commitments regarding mercury emissions made agreement and compromise between the United States and other nations on this issue unlikely. Yet the change of US administrations has led to a change in the role of US negotiators, demonstrating the significance of the

* General steps in the US negotiation process are mentioned here, but it should be noted that the details of how the United States internally decides negotiation positions and the roles of negotiators in any given negotiation is governed by the detailed requirements of the Circular 175, or C-175, process. This process involves, among other elements, the creation of a C-175 memo that gives negotiators very explicit parameters they are required to follow in negotiations. It also requires the transmittance from the executive branch to the US Senate Foreign Relations Committee of a ratification package with an explanatory memorandum on each provision of the final treaty text. For more information, see the US State Department overview of the C-175 process at http://www.state.gov/g/oes/rls/rpts/175/1265.htm.

US position to negotiations of multilateral environmental agreements. The United States is now interested in a meaningful mandatory reduction of anthropogenic mercury pollution. Hence, the role of the US delegation that will be attending the INC is to ensure that this happens in the most effective way possible, while working for a set of treaty provisions that the United States can feasibly implement within both the confines of its economic and domestic political situation and existing laws where possible. Such laws with regards to mercury pollution include the Clean Air Act, Clean Water Act, the Toxic Substances Control Act (TSCA), and the Federal Insecticide Fungicide and Rodenticide Act (FIFRA). Regarding the importance of negotiating a treaty that is congruent with domestic laws, it is worth noting that the United States has yet to ratify the POPs Convention, which the United States signed during the administration of President George W. Bush. The continued non-ratification of POPs by the United States is largely due to the fact that effective implementation of POPs requires amendments to the Toxic Substances Control Act (TSCA), a notoriously weak law. However, until Congress makes the necessary changes to TSCA the Senate will not give its advice and consent to the Stockholm Convention.

US negotiators can take several actions that will contribute to the success of the negotiations and resulting mercury treaty. First, the delegation must actively work with delegations from other countries outside of the regular meetings of the INC. Without such work it is unlikely that the United States will be able to have its objectives met as fully as possible, ensure that other key nations also participate in the treaty, and that the treaty enters into force. Second, it is also helpful if the United States designates a multilingual delegation. While a multilingual delegation is not necessary, it can help the delegation gain valuable knowledge and information from other delegations outside of the context of formal negotiation. Similar steps would very likely assist in the smooth negotiation of multilateral environmental treaties on other issues.

In addition to these steps that deal directly with the treaty and its provisions, experiences from other negotiations have demonstrated that the US delegation benefits from holding other side events less directly related to the negotiations. Examples of these events would be: meeting frequently with all relevant US stakeholders, especially including congressional staff, and hosting dinners and other informal meetings with key representatives of other countries outside of the normal working hours of the INC meetings. While such events do not necessarily contribute markedly to the substantive progress of negotiations, they enhance communication between delegations, making delegations more likely to be open to others' offers and objectives. This point is also made by Smith (2009, 149), who states that such nonplenary events also help to foster agreement between parties to the negotiations. Furthermore Smith (2009, 148–49) emphasizes that these events reduce the contention created by the repetition common to statements made in plenary. If it is to fulfill its objectives in the negotiations for the mercury treaty, the United States must work to overcome the current resistance to this treaty. Such resistance comes particularly from developing countries that, in the main, have contributed little to mercury pollution problems.

Through taking care to address matters such as the inclusion of delegates with multilingual ability, working outside of INC sessions to understand and include other nations' concerns, and through informal diplomatic methods, the United States will be in a much better position to emphasize to and convince resistant states that such a treaty is in their self-interest.

Furthermore the US delegation must develop the US position on any treaty in conjunction with Congress. Because of the central role of the US Congress in ratifying and implementing any treaty, the US delegation should always include members of the US Congress, or staff of the members of Congress, in the negotiation process for environmental treaties. This includes permitting and encouraging members of Congress or their staff to sit in on most negotiation discussions, informing key members of Congress about developments in the negotiations, and being receptive to their input. Without this heavy involvement of Congress in the negotiation process, it will be difficult for Congress to ratify the resulting treaty or to amend or pass legislation necessary to fulfill treaty obligations.

Another critical issue to which administrations have historically paid insufficient attention is the selection of a base year.* Whenever a base year is specified from which nations are to reduce the production of a substance, the relative economic, scientific, and political implications must be fully understood before agreeing with other nations. In the development of the Kyoto Protocol, American negotiators agreed to 1990 as the base year against which nations would reduce carbon dioxide ($CO_2$) emissions by an agreed percentage. However, the US economy grew dramatically in subsequent years. As a consequence of this growth the United States was unable to retreat to emission levels based on 1990 emissions without adverse economic consequences. This base year, though, proved very advantageous to the European Union (EU). Other nations agreed to treat the EU as one unit, freeing the EU to trade $CO_2$ emissions among member countries. Two of its larger nations, Germany and Great Britain, had emission reductions to spare owing to the reunification of Germany and subsequent closure of inefficient factories, and domestic political and economic factors that caused the decline of Great Britain's coal industry. Thus, the EU could easily allow some emissions increases to countries that were in a period of rapid economic growth and still meet its overall emission reductions as a single unit. Russia also found itself in an advantageous position to ratify the Kyoto Protocol, as the fall of the Soviet Union meant that inefficient facilities were no longer subsidized by the government and collapsed. Thus, in the case of the mercury treaty, or any subsequent environmental treaty, if a base year is proposed from which emissions or production levels of mercury would be reduced, US negotiators should understand what time period would result in the least economic burden and allow the United States to receive the most credit for its past unilateral actions. By doing so, the US delegation can help ensure that a key party to this treaty will be willing to ratify and implement it, in contrast with US actions regarding the Kyoto Protocol.

* In treaties that cover pollutant emissions, the base year is the year to which reductions of the emissions of the pollutants in question will be compared.

## IDEAL ELEMENTS OF INTERNATIONAL ACTION ON MERCURY

Provisions of the mercury treaty, as planned under the mandate provided by the Governing Council, will make significant contributions to controlling international anthropogenic mercury pollution. However, the inclusion of some additional provisions in the mercury treaty is warranted if the treaty is to provide the maximum environmental and health benefits internationally for the least cost possible. While several of the recommendations that follow are related to the control of heavy metals other than mercury, these recommendations are still highly relevant to the ongoing negotiation process. They directly follow from possibilities inherent in the current negotiations and are based on experiences with the development of other chemicals treaties.

### ADDING MECHANISM FOR OTHER HEAVY METALS

As mentioned above, it is vital to the international community and to the long-term success of a multilateral environmental agreement to have a sufficiently expansive scope. This should certainly be true of the final mercury treaty. First, heavy metals other than mercury should be regulated by the treaty. Discussions at the 25th Governing Council meeting that developed the mandate for this treaty also considered whether to regulate anthropogenic lead and cadmium pollution in this treaty (Appleton, 2009). However, despite the attempts of the countries in the EU and other nations to have other substances included for possible regulation under this treaty, the United States and others had references to the control of lead and cadmium removed from the final mandate. The United States refused to include other heavy metals in the mandate for the treaty despite the fact that it has ratified the Aarhus Protocol on Heavy Metals, which regulates long-range air transport of mercury, lead, cadmium, and other heavy metals on a regional basis (UNECE, 2008b). Additionally, the United States has, to date, maintained that the mercury treaty should have no adding mechanism for the possible future inclusion of other heavy metals. For the United States to have agreed to a binding regional agreement that includes other heavy metals but argue against an adding mechanism that would allow these same heavy metals to be regulated by the evolving international agreement is both illogical and inconsistent. The Governing Council decision at the 25th GC/GMEF regarding lead and cadmium noted that lead and cadmium continue to be of great concern and "that further action is needed to address the challenges posed by lead and cadmium" (Governing Council, 2009, 20). Additionally, an adding mechanism to include other heavy metals under the mercury treaty is theoretically possible with the mandate received from the Governing Council. The Governing Council mandate could perhaps also be expanded to explicitly include an adding mechanism (van Hoogstraten, 2009).

Negotiation of a mercury treaty thus provides an excellent opportunity to be able to control anthropogenic production, use, and effects of these other heavy metals. Adding mechanisms have been successfully implemented in other agreements, most notably the Stockholm Convention on Persistent Organic Pollutants. The POPs agreement initially contained twelve chemicals (the "dirty dozen") but included an adding mechanism for the addition of other persistent organic pollutants as identified by a scientifically based process. Were this to be done in the case of the mercury treaty, further years of health and environmental damage from these two metals would not occur, as they would if lead and cadmium are not managed in this agreement. In humans lead is a neurological toxicant and can also cause kidney damage, as well as a number of other health problems (US EPA, 2009a). Cadmium causes both lung and kidney disorders in humans, in addition to other health issues (US EPA, 2007). As with mercury, exposure to airborne lead and cadmium results in part from the burning of fossil fuels and waste incineration. Hence, these metals also need to be controlled internationally to protect human health, let alone ecosystem health and environmental quality.

With the POPs agreement, the adding mechanism is a tiered system. New chemicals to be controlled are added to one of three annexes, depending on their potential toxicity (Stockholm Convention, 2008b). When a chemical is added to the POPs treaty, the mechanisms for controlling that chemical are also specified according to a scientifically based process. Were a similar adding mechanism to be used for the mercury treaty, other toxic heavy metals, including lead and cadmium, could be added to the treaty as the international community was ready to regulate them globally. Nine chemicals have already been added to the POPs agreement since it was negotiated. Rather than undergoing the complications involved in negotiating a new treaty to control these chemicals, the international community simply had to initiate the established procedure through the POP Review Committee (Stockholm Convention, 2008c). An adding mechanism for the mercury treaty, or any other treaty to control a particular type of toxic substance, would provide the same advantages, and would allow for the addition to the treaty of other toxic heavy metals that are used, emitted, or further concentrated by human activities.

Disadvantages of an adding mechanism include that it may be politically difficult to negotiate. Some states may not want the additional legal responsibility of controlling their emissions of heavy metals that other states deem necessary to control—or other substances in the general case of other treaties. States may also have difficulty trusting that the technical bodies of conventions will make scientific rather than political decisions. Nonetheless, these disadvantages can be overcome. Resistant states can frequently be swayed by the positions and influence of key countries in the international arena. For example, were the United States and the EU to support an adding mechanism for the mercury treaty a number of countries would likely follow their lead. Developing countries would be especially likely to support an adding mechanism if developed countries offered additional technical support and capacity building to developing countries to comply with any additional requirements resulting from the adding mechanism. Thus, political compromises would be necessary to have such an adding mechanism become a part of the treaty. The necessity of compromise is particularly significant when considering how to include an adding mechanism while maintaining the political viability of the treaty. Legal requirements of an adding mechanism for a treaty include the creation of a body to accept and review applications for new substances, in this case metals, to be added to the treaty, and the establishment of the review process itself.

## Building on the Aarhus Protocol to the LRTAP Convention

Whenever possible new treaties should build on existing conventions, utilizing work already done regarding the issue and working to prevent fragmentation of the international regime surrounding the issue at hand. With regards to mercury and other heavy metals, the Aarhus Protocol to the LRTAP convention already places regionally based restrictions on mercury, lead, and cadmium pollution. Despite the long history of US opposition to a global treaty on mercury pollution, the United States has been party to the Aarhus Protocol since 1981 (UNECE, 2009). This treaty, though regional, sets an excellent precedent for the upcoming mercury treaty. It aims at regulating anthropogenic atmospheric emissions of mercury and thus limits international mercury pollution. It places restrictions on emissions from "industrial sources (iron and steel industry, non-ferrous metal industry), combustion processes (power generation, road transport) and waste incineration" (UNECE, 2008b), all of which will be targets of the new treaty. The Aarhus Protocol also outlines the optimal ways to control emissions from these sources. Therefore, it could help simplify the negotiation process and the work of the INC if the Aarhus Protocol to LRTAP were used as a partial model for the mercury treaty.

## Addition of Methylmercury to the Stockholm Convention

In addition to using existing treaties to help construct successful new treaties in a regime, actions possible through existing treaties regarding the issue at hand should be taken, or at least investigated. For the effective management of global mercury pollution, it would be highly advisable for the international community to add methylmercury to the lists of substances regulated under the Stockholm Convention on Persistent Organic Pollutants. Methylmercury likely qualifies as a persistent organic pollutant under the POPs convention because it is very toxic to humans and wildlife, is a carbon-based molecule, and bioaccumulates efficiently (US Geological Survey, 2009). The control methods of the POPs convention are also appropriate for controlling this organic pollutant. Furthermore, all the mechanisms are in place for the control of methylmercury were it to be added to the Stockholm Convention. While controlling mercury emissions will, over time, decrease levels of methylmercury in the environment, it would also be beneficial to control methylmercury through a separate channel. Doing so would recognize the particularly toxic nature of methylmercury and that it is ultimately present in the environment through different means than is elemental or inorganic mercury. Usually the mercury that is initially released through human activities is elemental mercury (US EPA, 2009b). Methylmercury is not introduced to the environment in significant quantities directly through human activities (US EPA, 2009b). Rather, elemental mercury is converted to methylmercury through biogeochemical processes that occur after an ecosystem has been contaminated with elemental mercury (US Geological Survey, 2009). Therefore, because of its unique chemical nature and mode of production when compared with elemental mercury, methylmercury should be managed under the Stockholm Convention in addition to the regulation of elemental mercury in the mercury treaty. Since the Stockholm

Convention is already in effect, international regulatory action on methylmercury could be expedited if this mechanism were used.

## RATIFICATION AND IMPLEMENTATION IN THE UNITED STATES

Unlike domestic legislation that automatically becomes law if passed, international treaties must be ratified and implemented by national governments. Without a sufficient number of ratifications the treaty cannot enter into force. Without implementation by national governments, the treaty and its secretariat do not protect the environment in the ways intended by the treaty. Treaties can stipulate actions to be taken against states that do not fulfill their treaty obligations, but international bodies cannot force national governments to fulfill these obligations. Thus, implementation is a key part of the treaty process. Understanding how it occurs is central to understanding global environmental negotiation and regulation. Implementation processes vary widely by country, so this section will use the United States as a case study. In addition to the fact that this is the case to which the authors have the most authority and ability to speak, the US ratification and implementation process has been the subject of much controversy in recent global environmental treaties. With the Kyoto Protocol on climate change, stark differences between the branch of the US government responsible for negotiation and the powers of the branch responsible for ratification caused an almost complete reversal of the US position on this protocol (Brunneé, 2004). When the United States does ratify a treaty it generally implements the provisions of that treaty through extant legislation, the development of implementing legislation, and regulations developed by executive branch agencies pursuant to such legislation.

### SIGNATURE OF A TREATY

For any environmental treaty, after INC meetings are completed and treaty text is finalized the next step for national governments is to sign the new treaty. Signature indicates a certain degree of agreement of the national government with the treaty and its provisions. Under the Vienna Convention on the Law of Treaties, nations that have not ratified but have signed treaties are expected to not significantly violate the terms of the treaty (van Hoogstraten, 2009). Though the United States has not ratified this Vienna Convention it still abides by many of its provisions, including largely abiding by the terms of treaties it has signed but not ratified (US Department of State, 2010; van Hoogstraten, 2009). In the US government the executive branch is responsible for the signature process. The US Department of State is the first entity involved in the decision on signature (and is also responsible for the negotiation of the treaty). Once the negotiations have concluded, the State Department is responsible for the main legal input to the president on whether to sign the treaty. If it advises the president to sign the treaty, one of two recommendations must be made. The first option is to certify that domestic laws exist under which the treaty can be implemented. If such laws do not exist, then the State Department must identify what existing laws or regulations must be amended to meet treaty obligations. With these recommendations in mind the president then has the option of signing the treaty. Thus, US negotiators can best assure that the president will sign a treaty by working for provisions that can be implemented under existing US law, or with the least changes necessary to existing law to allow the effective implementation of the treaty obligations.

## Ratification of a Treaty

If the president signs a treaty it will normally be submitted to the US Senate for ratification. In practice, treaties have generally not been sent to the Senate for ratification until all necessary implementing legislation and regulation has been put into place. As stipulated in Article II, Section 2 of the US Constitution, the treaty will have to be approved by two-thirds of the Senate in order for it to be ratified. Once it has been ratified by the US Senate, an instrument of ratification is then deposited (placed on record) with the depositary of the treaty (van Hoogstraten, 2009).

Division of responsibility and power between the two distinct branches of the US government presents a unique challenge to the ratification of a treaty by the United States. Because the Senate is not necessarily composed of a two-thirds majority of either party, and because the Senate can be of a different party than the president, ratification of a treaty can be much more difficult in the United States than in the parliamentary systems common in most countries. With many recent environmental treaties, the Senate has blocked ratification of a treaty that the president signed. Though the George H.W. Bush administration signed the Convention on Biological Diversity, the Senate refused to ratify the treaty (DeSombre, 2005, 195). The United States also has not ratified the POPs Convention, mostly because, as discussed previously, implementation of the convention will require amending the Toxic Substances Control Act (TSCA), something the US chemical industry has resisted since the inception of the law in October 1976. Furthermore, amendment of TSCA would require action by both the House and Senate. Perhaps the most infamous case of Senate refusal to ratify a treaty is the case of the Kyoto Protocol. The Clinton administration backed the protocol fully. However, the administration failed to gain the support of Congress for regulation of greenhouse gas emissions (Smith, 2009, 152). Even before the negotiations had officially concluded, the Senate voted 95-0 in favor of the 1997 bipartisan Byrd–Hagel resolution (S. Res. 98) (Brunneé, 2004). This resolution stated explicitly that even if the Clinton administration submitted the Kyoto Protocol to the Senate, the Senate would not ratify the treaty (Brunneé, 2004). Therefore, the Clinton administration never submitted the protocol to the Senate for ratification after signature.

How can the same fate be avoided for the mercury treaty, or even for subsequent environmental treaties? One way to avoid these problems is to avoid language that contradicts established laws and regulations. Beyond this, past experience, particularly with the Byrd–Hagel resolution, shows that close work between the executive branch and the Senate is critically necessary to the success of a treaty. Where existing legislation needs to be amended or new laws enacted, close cooperation with the House is also critical. Hence, US negotiators, and the Executive Branch more generally, must learn from the recent past and actively engage the Senate in the negotiation process, in addition to ensuring the provisions are equitable among major trading partners. Engaging Congress more fully should include timely briefings of relevant congressional committees prior to the INC and other important meetings and debriefings following such meetings. Members of Congress and congressional staff should also be included as observers on US delegations when appropriate.

## Implementation of a Treaty

Once a treaty has been ratified, its provisions carry the force of law and must be implemented as such. Article VI of the US Constitution states that "all Treaties made, or which shall be made, under the Authority of the United States, shall be the supreme Law of the Land." Frequently, treaties bind governments to, for example, target reductions in emissions of a pollutant by a certain date. Yet these treaties often do not specify the methods by which the targets are to be met in a manner specific enough to allow for domestic enforcement of the treaty provisions without additional legislation or regulation. For these reasons, in modern times, the United States has avoided "self-executing" treaties. Instead, implementing legislation or amendments to existing legislation are developed to specify how treaty obligations will be fulfilled.

With regards to mercury, some existing legislation and regulation would provide for a certain degree of control of US anthropogenic mercury emissions. The Mercury Export Ban Act of 2008 placed an immediate ban on the trade or transfer of mercury between federal agencies and outlawed the export of elemental mercury from the United States by January 1, 2013 (Mercury Export Ban, 2008); it was enacted even though the United States is a net exporter of mercury (Mercury Export Ban, 2008). Earlier legislation such as the Mercury-Containing and Rechargeable Battery Management Act, which began phasing out mercury in batteries in 1996, has also contributed to domestic decreases in the use and emission of mercury (Mercury-Containing Battery Act, 1996; US EPA, 2009b). By 2007 domestic regulation of waste incinerators had reduced total mercury emissions by 45% in nine years (Selin, 2005). EPA rules regarding the emission of mercury include both interpretations of long-standing, broader laws regarding pollution (such as the Clean Air Act) and mercury-specific rules (US EPA, 2009). The EPA's rulings on mercury emissions set limits for emissions from iron and steel foundries, battery disposal, chlor-alkali plants, and incinerators as well as established limits for mercury loads in water (US EPA, 2009).

As mentioned earlier it is desirable to negotiate provisions that require as little legislative change as possible for implementation. Despite these preexisting regulations of mercury, the lack of control on mercury emissions from coal-fired power plants is one area where US management of anthropogenic mercury emissions is lacking. Additionally, the mercury treaty will have provisions related to technical support and capacity building for developing countries. The Clean Air Act likely provides the necessary framework for the United States to regulate mercury emissions, but further steps may be needed for the regulatory process to take place. Congressional action will probably be needed on budgetary matters regarding US foreign aid on issues of technical assistance and capacity building. A further consideration is the implementation of recommendations for the treaty that were presented earlier. If an adding mechanism were included, domestic controls on heavy-metal pollution would need to be changed depending on the specific substances, targets, and pollution sources that would be regulated. Finally, methylmercury is largely unregulated in the United States as a separate compound from elemental mercury. Therefore, while the United States does have a good baseline set of laws and regulations governing anthropogenic emissions of mercury, amendments to the existing body of laws may be necessary to most effectively implement all the provisions of the mercury treaty, should it be adopted with all the recommendations contained herein.

## CONCLUSION

For the mercury treaty, the change in US position at the 2009 UNEP Governing Council was of vital importance. US support of the mercury treaty has been and continues to be crucial to the process, from the mandate through the working group and INC meetings, to the entry into force and implementation of the treaty. Without the support of the United States its major trading partners, including China and Brazil, would be much more reluctant to move forward with the mercury treaty. Thus, the change in the US position is significant not only because the United States itself is a significant emitter of mercury, but because other nations that have a considerable impact on anthropogenic mercury pollution will now also participate in the treaty and its negotiation.

Acknowledging the importance of the change in the US position, further refinement of its position is necessary for the treaty to be as effective and efficient as possible. The United States should advocate the inclusion of an adding mechanism. Such a position is logical, as outlined previously, because it is the most time- and resource-efficient way of controlling anthropogenic production and concentration of other heavy-metal pollutants. The negotiation of the mercury treaty is an excellent opportunity for the international community to develop a way of minimizing the negative effects of heavy-metal pollution. However, the adding mechanism that would allow for this will not be a part of the mercury treaty without the full support of the United States. Therefore, it is hoped that the US administration will modify its previous statements regarding such an adding mechanism.

Beyond the US role in the mercury treaty, the UNEP secretariat has clearly gained valuable experience from its role in the preparation for and negotiation of previous international chemical agreements such as POPs and the Rotterdam Convention. The results of the Open-Ended Working Group demonstrate that the secretariat accomplished what was necessary to expedite substantive discussions prior to the OEWG meeting. In addition, previous chemicals treaty negotiations have demonstrated that a strong chair is critical for the success of negotiations and the preparatory process leading up to the INC. To effectively fulfill their roles chairs need to be capable of listening to all sides as a negotiation progresses, determining the appropriate time to draft initial negotiating texts, and then creating from that strong and workable compromise texts as discussions continue. Furthermore, chairs must be able to use the bureau effectively to help with developing solutions to conflicts during the negotiations, and then present those solutions to the full INC through the bureau. It appears from the OEWG results that John Roberts (UK), the chair of that meeting, has the ability to guide the negotiations so that they proceed smoothly and productively should he be chosen chair of the INC. Therefore, the international community can be confident that the negotiation of the new global and legally binding treaty on anthropogenic mercury pollution will progress effectively and likely produce an outcome that will effectively control anthropogenic mercury pollution. However, some very feasible changes to the currently planned outcome, especially an adding mechanism for other heavy metals, would greatly improve the possible accomplishments of the negotiations that will be completed in 2013.

Finally, the mercury treaty provides an opportunity for those involved in its negotiation to synthesize and act on the lessons learned from previous negotiation, ratification, and implementation experiences. In a new era of US cooperation on international environmental protection, the treaty will also provide an interesting case study in global environmental politics. Depending on its final provisions, the mercury treaty may represent a new turning point in multilateral environmental agreements, and will certainly serve as a significant example for environmental treaties yet to be developed.

## APPENDIX I: GLOSSARY OF ACRONYMS

$CO_2$: Carbon dioxide, a common greenhouse gas
(US) EPA: (United States) Environmental Protection Agency
EU: European Union
FIFRA: Federal Insecticide, Fungicide, and Rodenticide Act
GC/GMEF: Governing Council/Global Ministerial Environment Forum
INC: Intergovernmental Negotiating Committee
LRTAP: The Convention on Long-Range Transboundary Air Pollution
NGO: Non-Governmental Organization
OEWG: Open-Ended Working Group
PIC: (Rotterdam Convention on) Prior Informed Consent
POP: (Stockholm Convention on) Persistent Organic Pollutants
TSCA: Toxic Substances Control Act
UNECE: United Nations Economic Commission for Europe
UNEP: United Nations Environment Programme

## APPENDIX II: UNITED STATES PROPOSES DECISION ON MERCURY AT 25TH GC/GMEF

Full text of Reifsnyder, 2009:

Statement of Daniel A. Reifsnyder, Deputy Assistant Secretary for Environment and Sustainable Development U.S. Department of State

UNEP GC 25, February 16, 2009, Nairobi, Kenya

### Intervention on Mercury

Thank you Mr. Chairman, I greatly appreciate the opportunity to express the views of the United States on this important issue, and I encourage other delegates to take this opportunity in the plenary to express their own views so we can carefully gauge the progress we may be able to make at this meeting prior to breaking into a contact group.

### Implications of Mercury Pollution

The United States has made great strides in addressing mercury contamination at home.

However, more than half of all mercury deposition within the United States comes from sources outside our borders. In the United States, mercury is the most common cause of fish contaminant advisories, and presents a major concern for public health as well as fisheries and commerce. The US Environmental Protection Agency (EPA) and Food and Drug Administration (FDA) warn that pregnant women and young children should avoid consuming fish that are high in mercury. Despite the major efforts currently undertaken domestically by U.S. Federal, State, and local authorities to reduce mercury levels in rivers and streams, currently all 50 states have fish advisories. In recognition of the importance of this issue, in 2008, under the leadership of our now President Obama, the United States' Congress passed the Mercury Export Ban Act to ban exports of elemental mercury beginning in 2013. However, neither the United States, nor any other country can achieve sufficient reductions of mercury risks to protect the health of its citizens without serious cooperation internationally to reduce global mercury emissions.

### International Discussions on Mercury

I believe it is valuable to recall how the Governing Council has arrived at this point in the discussion. In 2001, the international environmental community began to focus on mercury when it launched the global mercury assessment to understand better the significance and sources of mercury as a global pollutant. The report from that assessment clearly set out the threat that mercury poses to human health and the environment, and the nature of mercury as a global pollutant. In 2003 and 2005, the Governing Council began substantively to address concerns over mercury, first by establishing a Mercury Program in UNEP, and second by launching a mercury partnership program. The United States has been a strong supporter of both of these developments and has contributed over \$5 million to the Partnership. We are pleased with the significant work done by UNEP and the UNEP Global Mercury Partnership.

Over the years, there has been a significant amount of highly productive and valuable work that has further clarified our knowledge of the issues surrounding mercury, especially as they relate to anthropogenic sources of mercury and its long-range transport.

The productive discussions at the two meetings of the Open-Ended Working Group, the numerous meetings on partnerships, and the intergovernmental process to develop UNEP's global mercury assessment have provided vital information to evolve our understanding since the early days of discussion of this issue. We congratulate Achim Steiner and the UNEP Chemicals Secretariat for the work that has enabled us to explore this issue in depth and further consider the strengths and weaknesses of various approaches to address mercury globally.

## The U.S. position on mercury

We have now arrived at a point where there is a call to come together to launch an Intergovernmental Negotiating Committee to develop an international agreement on mercury;

The United States now joins that call.

We are prepared, Mr. Co-Chair to help lead in developing a global legally binding instrument for mercury. We believe that:

- — Now is the time for governments to launch an Intergovernmental Negotiating Committee (INC);
- — The first negotiating session should begin this year with the goal of completing negotiations prior to the 2012 Governing Council (GC)/Global Ministerial Environment Forum (GMEF);
- — The mandate of the INC should be devoted exclusively to mercury;
- — It should be comprehensive, addressing all significant sources of mercury emissions, with particular attention to sectors that have the greatest global impact such as coal-fired power plants and other sources of unintentional air emissions;
- — Financial resources for implementation should focus on priority issues of greatest global concern;
- — It should include approaches tailored to specific emissions sectors, and contain a level of flexibility to achieve our global goals while allowing countries discretion in terms of their path to implementation;
- — Governments should support the UNEP Mercury Program and Global Mercury Partnership to continue their work concurrent with the negotiations

To aid in understanding the specific approach we advocate, my delegation has submitted a conference room paper that urges a simple streamlined approach to a decision at this meeting.

It is clear that mercury is the most important global chemical issue facing us today that calls for immediate action. Mercury is a chemical of global concern specifically due to its long range environmental transport, its persistence in the environment once introduced, its ability to bio-accumulate in ecosystems, and its significant negative effects or human health and the environment. The United States does not support adding additional substances to an agreement on mercury, or diverting valuable time and attention to other issues by debating criteria and parameters for an adding mechanism.

We urge delegates to focus on those issues where we can find agreement.

This is not to say that we do not understand concerns that colleagues have variously expressed about other chemicals, in particular lead and cadmium. In our view such concerns are better addressed in other fora such as Strategic Approach to International Chemicals Management (SAICM), which will meet later this year. In particular, we are strongly interested in considering a partnership to remove lead from paint, similar, perhaps, to the partnership for Clean Fuels and Vehicles that has so successfully helped remove lead from gasoline. However, we do not favor expanding the mandate on mercury to other chemicals.

We should seek to ensure broad participation among governments in a mercury agreement by including flexibility such as transition periods and phased implementation. We should apply tailored approaches that address the sector-specific nature of mercury emissions and the technologies and methods used to reduce emissions. We need to be creative and flexible in our approaches, while at the same time ensuring that we make significant progress.

The United States supports further action by UNEP in the near term to access options for reducing emissions from coal combustion and other significant sectors, and comparing the relative costs of different options. It will also be vital to understand better the size and types of sources that need to be addressed, as well as the existing pollution control equipment currently in place. This type of survey should include information on the co-benefits of efforts to address conventional pollution control and climate change. This work will help to inform the INC to understand better the options for this sector and prioritize actions to reduce mercury emissions.

The timing of implementation of commitments for different sectors may vary depending on the complexity of the actions, but it should be clear that all significant sources of mercury should be included in the mandate of the INC.

### Financial Consideration

We realize that there is much interest and need to discuss the financial aspects of an agreement and we understand that this is an important issue for many countries. In order to concentrate available financial resources on issues of greatest global concern, our focus should be on efficient, cost-effective actions to address priority sources that cause the most environmental impact. Awareness building programs are important components of capacity building in developing countries and should also be supported. The UNEP Global Mercury Partnership has been very effective in raising awareness of mercury issues among a broad range of participants and can serve as a model for future action.

In fact, regardless of the outcome of the decision on the mercury instrument, it is imperative that the Partnership activities continue and expand, since any agreement will take years to come to fruition.

Under the guidance of the Partnership Advisory Group, the Partnership can play an important role for near term action on mercury to priority sectors. We urge all donors, particularly those who have not done so, to contribute to the Partnership as an effective near team instrument to address mercury pollution.

While we strongly encourage UNEP to continue to support the work of the partnership and to take the lead on conducting additional activities during the interim period, it is imperative that any request that we make of UNEP include explicit understanding that the work will be supported with appropriate levels of funding to complete the task. Delegates should work together to ensure that we don't overburden UNEP in this regard, so that it can continue to function as an effective facilitator of our efforts.

We have an opportunity at this GC to work together on a decision that will allow us to take further steps in addressing global mercury pollution. This delegation is prepared to engage fully with other governments and stakeholders to seek solutions that ensure a comprehensive path forward that effectively addresses mercury emissions on a global basis. I thank you Mr. Co-Chair for this opportunity to address the Council, and anticipate that productive and positive discussions will take place at this meeting that will set us on a path of cooperation and progress.

## REFERENCES


Appleton, Asheline, Melanie Ashton, Tallash Kantai, Wangu Mwangi, Keith Ripley, and Andrey Vavilov, PhD. 2009. Summary of the 25th session of the UNEP Governing Council/ Global Ministerial Environment Forum: February 16–20, 2009. *Earth Negotiations Bulletin,* 16 (78).

Appleton, Asheline, Nienke Beintema, Xenya Cherny Scanlon, Leonie Gordon, and Andrey Vavilov PhD. 2007. 24th session of the Governing Council of the United Nations Environment Programme: 5–9 February 2007. *Earth Negotiations Bulletin,* 16 (60).

Ashton, Melanie. 2009. Briefing note on the mercury OEWG. *International Institute for Sustainable Development,* (October 26).

Ashton, Melanie, Tallash Kantai, Jessica Templeton, and Kunbao Xia. 2010. First meeting of the intergovernmental negotiating committee to prepare a global legally binding instrument on mercury: June 7–11, 2010. *Earth Negotiations Bulletin,* 28 (6).

Bai, Changbo, Paula Barrios, Maria Larsson Ortino, Richard Sherman, Andrey Vavilov PhD, and Kunbao Xia. 2005. 23rd session of the Governing Council of the United Nations Environment Programme: February 21–25, 2005. *Earth Negotiations Bulletin,* 16 (47).

Brunnée, Jutta. 2004. The United States and international environmental law: Living with an elephant. *European Journal of International Law,* 15:617–649.

DeSombre, Elizabeth. 2005. Understanding United States unilateralism: Domestic sources of U.S. international environmental policy. In *The Global Environment: Institutions, Law, and Policy*, Regina S. Axelrod, David Downie, and Norman J. Vig, eds. Washington: CQ Press.

Ganzleben, Catherine, Richard Sherman, Chris Spence, Andrey Vavilov PhD, and Hugh Wilkins. 2003. 22nd session of the Governing Council of the United Nations Environment Programme: February 3–7, 2003. *Earth Negotiations Bulletin,* 16 (30).

Governing Council of the United Nations Environment Programme. 2009. *Proceedings of the Governing Council/Global Ministerial Environment Forum at its twenty-fifth session.* United Nations, UNEP/GC.25/17.

Mercury-Containing and Rechargeable Battery Management Act. 1996. 104th U.S. Congress. Public Law 104–142.

Mercury Export Ban Act of 2008. 2008. 110th U.S. Congress. Public Law 110–414.

Reifsnyder, Daniel A. 2009. Statement of Daniel A. Reifsnyder, Deputy Assistant Secretary for Environment and Sustainable Development U.S. Department of State. Intervention on mercury. Paper presented at Opening Session of the Twenty-Fifth Governing Council/ Global Ministerial Environment Forum of the United Nations Environment Programme, Nairobi, Kenya, http://www.unep.org/Documents.Multilingual/Default.asp?DocumentID=566&ArticleID=6083&l=en&t=long.

Selin, Noelle Eckley. 2005. Mercury rising. *Environment,* 47 (1):22–35.

Smith, Richard J. 2009. *Negotiating Environment and Science: An Insider's View of International Agreements, from Driftnets to the Space Station.* Washington: Resources for the Future.

Stockholm Convention on Persistent Organic Pollutants. 2008a. About the POPRC. (cited January 14, 2010). Available from http://chm.pops.int/Convention/POPsReviewCommittee/AboutPOPRC/tabid/221/language/en-US/Default.aspx.

Stockholm Convention on Persistent Organic Pollutants. 2008b. Chemicals review process. (cited January 14, 2010). Available from http://chm.pops.int/Convention/POPsReviewCommittee/Chemicals/tabid/243/language/en-US/Default.aspx.

Stockholm Convention on Persistent Organic Pollutants. 2008c. What are POPs? (cited January 14, 2010). Available from http://chm.pops.int/Convention/ThePOPs/tabid/673/language/en-US/Default.aspx.

US Department of State. 2010. Vienna Convention on the Law of Treaties. (cited February 20, 2010). Available from http://www.state.gov/s/l/treaty/faqs/70139.htm.

US Environmental Protection Agency. 2007a. Technology transfer network air toxics web site: Cadmium compounds. (cited January 31, 2010). Available from http://www.epa.gov/ttn/atw/hlthef/cadmium.html.

US Environmental Protection Agency. 2007b. Technology transfer network air toxics web site: Mercury compounds. (cited February 2, 2010). Available from http://www.epa.gov/ttn/atw/hlthef/mercury.html.

US Environmental Protection Agency. 2009a. Lead in air: Health and environment. (cited January 31, 2010). Available from http://epa.gov/air/lead/health.html.

US Environmental Protection Agency. 2009b. Lead in air: Regulatory actions. (cited February 3, 2010). Available from http://epa.gov/air/lead/actions.html.

US Environmental Protection Agency. 2009c. Mercury: Laws and regulations. (cited April 16, 2009). Available from http://www.epa.gov/mercury/regs.htm.

US Geological Survey. 2009. Toxic substances hydrology program: Methylmercury. (cited February 2, 2010). Available from http://toxics.usgs.gov/definitions/methylmercury.html.

United Nations Economic Commission for Europe (UNECE). 2008a. Convention on Long-Range Transboundary Air Pollution: The 1979 Geneva Convention on Long-Range Transboundary Air Pollution. (cited February 2, 2010). Available from http://www.unece.org/env/lrtap/lrtap_h1.htm.

United Nations Economic Commission for Europe (UNECE). 2008b. Convention on Long-Range Transboundary Air Pollution: Protocol on Heavy Metals. (cited February 2, 2010). Available from http://www.unece.org/env/lrtap/hm_h1.htm.

United Nations Economic Commission for Europe (UNECE). 2009. Convention on Long-Range Transboundary Air Pollution: Status of ratification of the 1979 Geneva Convention on Long-Range Transboundary Air Pollution as of 2 September 2009. (cited January 19, 2010). Available from http://www.unece.org/env/lrtap/status/lrtap_st.htm.

United Nations Environment Programme (UNEP). 2009a. Mercury programme partnership: Summarized table of current partners. (cited January 12, 2010). Available from http://www.chem.unep.ch/MERCURY/Sector-Specific-Information/Docs/Indication%20of%20interest/PARTNERSHIP_Status_Table_September%202009_www.pdf.

United Nations Environment Programme (UNEP). 2009b. Overarching framework: UNEP Global Mercury Partnership. (cited January 4, 2010). Available from http://www.chem.unep.ch/mercury/Sector-Specific-Information/Docs/Overarching%20Framework.pdf.

United Nations Environment Programme (UNEP). 2009c. Relevant issues being considered in international forums and their possible impact on the mercury negotiation process. UNEP(DTIE)/Hg/WG.Prep/1/7.

United Nations Environment Programme (UNEP). 2010. Progress in the preparation of the study called for in paragraph 29 of Governing Council decision 25/5. UNEP(DTIE)/Hg/INC.1/15.

van Hoogstraten, David J. Former senior lawyer and policy maker at the U.S. Department of State, White House Council on Environmental Quality and the Environmental Protection Agency. 2010. Personal communication. February 15, 2010.

# Afterword: Through Your Eyes

*Li Xiong and Sharon L. Zuber*

## CONTENTS



> Mercury is the only heavy metal that has three states—solid, liquid, and gaseous—at ambient temperatures. This uniqueness makes it possible for mercury to be distributed globally.
>
> **—Li Xiong**

## INTRODUCTION

Woven throughout this book is the theme that mercury, a unique, attractive, poisonous element, has been spread throughout our globe by human activity. As Elizabeth Malcolm argues, "Compared to any natural process, human actions have greatly accelerated movement of this metal out of the earth's crust, where it was stable and sequestered, and transferred it to air, soil, and surface waters where it finds its way into aquatic food webs and, ultimately, into humans." If we are to understand the risks involved with using mercury, we must be able to share research, coordinate regulations, and communicate across disciplinary and geographic boundaries. Thus, this book explores mercury pollution with methods from and perspectives of different disciplines—themselves cultures—in a way that we hope is accessible to a wide audience.

## INTERDISCIPLINARY

Because mercury in most forms is invisible or hidden, people may not realize how ubiquitous it is in our everyday environment and how potentially hazardous. A Stanford University flyer advising replacing mercury thermometers warns that at room temperature a broken thermometer will emit "vapors that are colorless, odorless, tasteless, and toxic" and that "0.5 grams of Mercury from one broken thermometer can detectably pollute 5 million gallons of San Francisco Bay water" (Replace). The old mercury-filled thermometers are being replaced, but mercury may still be found in household thermostats, some batteries, and even dental fillings. As reported by *CNN News*, just as this book is going to press, the Food and Drug Administration (FDA) advisory committee recommends that FDA reexamine its 2009 findings that mercury amalgam dental fillings are not harmful, a stance still supported by the American Dental Association (Willingham, 2010). In "Mercury by the Numbers," Mike Newman and Kenneth Leung note that "making decisions about environmental issues are particularly difficult, but nonetheless, very important to our well-being. Environmental issues permeate our laws, politics, economics, personal health, and collective ethos." The necessity of interdisciplinary collaboration is vital.

The greatest exposure to mercury by humans around the world is from food, especially fish. Some indigenous peoples are most at risk if their diets depend upon the protein from the fish they catch in mercury-polluted waters. For these communities fishing may also be a ritual tied to religious and social practices, important cultural traditions. In the United States, FDA and the Environmental Protection Agency (EPA) advisories have caught media attention and found advocates in political figures such as Robert Kennedy Jr. In the study of *USA Today* articles between 2000 and 2009, Christine Mowery and Sarah Jane Brubaker calculate that the greatest number of articles on mercury pollution, in total, were about fish consumption, followed by articles about coal-fired power plants, electronics disposal, and vaccines. Also using sociological analysis, Kelly Joyce focuses on how FDA and EPA "shape the risks associated with methylmercury." However, the numbers can be confusing as can the information obtained from various Internet sources. John Drummond in "Input/Output" gives tools to investigate these sources, emphasizing how important it is that individuals read critically—including this book—and become informed consumers.

What brought mercury to the attention of the world was the Minamata disaster. Mentioned by several authors, this industrial tragedy that poisoned fish in Japan's Minamata Bay resulted in deaths and deformities. The struggle to determine the cause of what became known as Minamata Disease continues even after years of political, economic, and scientific investigation. Clearly, this struggle is a case "When the Scientific Vantage Is Not Enough." It took, in part, the still photography of W. Eugene and Aileen M. Smith, and as Justin Jesty contends, documentary films by Tsuchimoto Noriaki, to "Make Mercury Visible." Artists rendering their concern for the environment through theatrical productions, paintings, photography, and sculpture bring the subject of mercury into the world beyond the scientific laboratories, business board rooms, and halls of government. Reacting to details of the Great Depression hat-making industry, Kira Obolensky portrays on stage "a collective madness," "a vivid example of madness connected to the bargain we make with prosperity." Elizabeth Mead creates and teaches art "conscious of the impact the materials and processes" have on the environment. No longer are the effects of mercury pollution so isolated and invisible.

## INTERNATIONAL

Ultimately, finding solutions to the problem of mercury pollution, once exposed, will require collaboration across international borders. As Pep Fuller argues so convincingly:

> Domestic regulation alone cannot control mercury pollution within a state's borders because of the quantity of mercury released into the atmosphere and hydrosphere

> globally, and the effectiveness with which it and its metabolites (e.g. methylmercury) bioaccumulate and are transported for long distances in these systems. Mercury released into the atmosphere from coal burning in China is deposited onto U.S. soil and into U.S. waters. Mercury released from gold mining activities in South America and chlor-alkali plants in the U.S. eventually drains to the oceans, where it bioaccumulates to dangerous levels in larger fish such as tuna, consumed worldwide. Hence, mercury pollution does not respect national boundaries and is a problem that affects the global commons. With regards to the U.S. specifically, over half of the anthropogenic mercury in U.S. ecosystems is from international sources. (Reifsnyder 2009)

Daniel Jacob puts it this way: "We're all breathing each other's air." Jacob, a Harvard professor of atmospheric chemistry, was a chief researcher in a study of transcontinental air pollution in which "he traced a plume of polluted air from Asia to a point over New England, where samples revealed that chemicals in it had come from China" (Pottinger 2004). Pep Fuller's examples make clear that global, but specifically China–US cooperation, will be an important part of any solutions, not just to the problem of mercury poisoning but to all global pollution issues.

The title for this chapter emerged from one of our many conversations about the importance of being able to see issues from another cultural perspective. Too often we can only see a problem from a limited perspective constructed by our own ethnocentrism. It is too easy to blame countries such as China and then overlook other countries' contributions to the mercury pollution problem. For example, the United States recycles and has reduced its use of mercury; however, the recycled mercury finds its way to countries where artisanal miners may use the liquid toxicant, putting their health at risk, for extracting precious metals, metals that are crafted into expensive jewelry sold in the United States. We all must take responsibility for helping solve mercury pollution as well as other environmental problems, but to cooperate internationally we have to dispel fears and move beyond stereotypes.

In the early stages of research for the mercury Global Inquiry Group (GIG), people kept referring to China as "the world's largest source of nonnatural emissions of mercury" (Pottinger, 2004). Although some scholars noted that China was developing industrially at a similar pace to the United States thirty to fifty years ago, the questions loomed: Is there some way to skip over the stage characterized by thoughtless chemical and heavy-metal pollution? How can the world's developing countries learn from mistakes the United States and other first world countries have made?

Mercury has been a part of Chinese culture for centuries. Kris Lane documents that mercury "was not always feared and hated. Indeed, throughout most of human history, this ambivalent, seemingly luminal substance and its compounds were highly prized." It has been "found in royal tombs ... either to act as a preservative for the bodies or as a deterrent to would-be tomb raiders" (Quingbo, 2007). In the Forbidden City at the center of Beijing, the bright red pigment on the ancient buildings would have been achieved using mercury-laden cinnabar. Of concern today is that "Mercury and other pollutants from China's more than 2,000 coal-fired power plants soar high into the atmosphere and around the globe on what has become a transcontinental conveyor belt of bad air. North America and Europe add their own dirty loads to the belt. But Asia, pulsating with the economic rebirth of China and

**FIGURE A.1** Coal train from Yellow Crane Tower, Wuhan, China. Photo by Michael C. Newman, 2008.

India, is the largest contributor" (Pottinger, 2004). Standing at the top of the Yellow Crane Tower overlooking the Han and the Yangtse rivers in Wuhan, a visitor might see a long string of coal cars pass below (Figure A.1); they are similar to the many coal trains that pass through Williamsburg, Virginia, from the mountains of West

Virginia on their way to the Newport News Shipyard. If we look beyond the cultural differences, our similarities may provide spaces to connect and collaborate.

Yet even collaboration will necessitate tough decisions. Three Gorges Dam, the world's largest hydroelectric power source and China's engineering pride, represents a cleaner alternative to coal. However, the equation balancing coal versus hydroelectric power is not as simple as it seems. Building Three Gorges Dam and the consequent flooding of the gorges has resulted in loss of homes, sacred places, and habitat; water rising above places that had contained small coal mining facilities as well as burial caves creates an environment ripe for making methylmercury. How can we find a balance between these negatives with the intended preserving of life and land from flooding downstream, the need for fewer coal-fired power plants to reduce atmospheric pollution, and pride in a great engineering accomplishment?

At the same time that a DVD of Al Gore's *An Inconvenient Truth* cannot be bought in China, the country is responding in positive ways to environmental issues. In its preparation for the 2008 Olympics, recycling bins were placed next to trash receptacles along the streets of Beijing. In its effort to promote international economic opportunities, China seemed sensitive about its actions on the global stage and visibly responsive to some environmental concerns. On June 1, 2008, China passed a law requiring that each person pay for plastic bags at stores as a way to reduce pollution and plastic waste. In contrast, only in June of 2010 did a single US state, California, pass legislation "prohibiting pharmacies and grocery, liquor and convenience stores from giving out plastic bags" (Young, 2010). Also, in major Chinese hotels, the keycard serves to conserve power use in every room; when no one is in the room the power is turned off until reactivated by the card. The contradictions between these energy-saving measures and the increasing harmful airborne mercury pollution coming from within China suggest the uneven movement we face toward cleaning up our earth. When companies and governments are responsible for monitoring and setting standards, obstacles abound, including conflicting scientific research about what are the best solutions. Perhaps we can take inspiration from the 2008 Olympic theme "One World, One Dream" to make stronger commitments to reduce mercury pollution, commitments that unite us across borders.

## CONCLUSION

Sydney Plum, environmental writer and advocate for the loons and eagles threatened by mercury-polluted lakes in New England, asks, "Why did I think that a poem would change the behavior of my neighbors?" We now ask, "What can the words in these chapters do to change the way our world responds to mercury pollution?" Rachel Carson and Lewis Thomas created metaphors that changed the way the world thought. Carson, a biologist, imagined a "silent spring" for her audience of nonscientists. Thomas, a medical doctor, connected the micro with the macro by comparing a human cell to the earth. His connection between human biology and the earth implies a connection both among all humans and between humans and our living universe. We hope that something in these pages will inspire readers to think in new ways about mercury pollution. By making connections across disciplinary and national cultures, perhaps we can remember our similarities as we face conflicts based on our many differences, conflicts that threaten the environment and the quality of human life.

## REFERENCES


Carson, Rachel. 1962. *Silent Spring*. New York: Houghton Mifflin.

Pottinger, Matt. 2004. Invisible export—a hidden cost of China's growth: Mercury migration. *Wall Street Journal,* Dec. 20, 2004.

Replace your mercury thermometers BEFORE they break! www.stanford.edu/dept/EHS.

Quingbo, Duan. 2007. Scientific studies of the high level of mercury in Qin Shihuangdi's tomb. Appendix in *The First Emperor: China's Terracotta Army*. Jane Portal and Hiromi Kinoshita, eds. UK: British Museum Press.

Thomas, Lewis. 1974. *The Lives of a Cell: Notes of a Biology Watcher*. New York: Viking Press.

Willingham, Val. 2010. Panel urges FDA to take another look at mercury dental fillings. *CNN News,* December 15.

Young, Samantha. 2010. Plastic bag ban: California moves to ban plastic bags at grocery stores. *Huffington Post*, Aug. 18, 2010.

# Index

Page numbers followed by *f* or *t* refer to figures or tables.

## D



## E



## F

## N

## X



## Y